THE DISTANCE BETWEEN ZERO AND ONE

LOVE, LIGHT, AND THE MATHEMATICS OF EVERYTHING

KEPA

CONTENTS

FOREWORD

You are holding a love letter.

Not from me to you — though there is love in these pages — but from the universe to itself, written through these hands, on these pages, in this moment that somehow called both of us here across the vast improbability of everything that had to happen for this sentence to exist and your eyes to read it.

Feel that before you think about it.

Your eyes are moving across small marks. Dark shapes against a pale field. Light strikes the page, reflects into your pupils, lands on the retina, and becomes signal. Your brain takes those signals and does what it has learned to do across a lifetime of language: it turns pattern into meaning. Ink becomes sound inside the mind. Sound becomes image. Image becomes recognition. Recognition becomes feeling.

That is not ordinary.

That is the universe using matter to speak.

A book is a machine made of stillness. It sits there quietly until a living nervous system enters the circuit. Then the machine wakes up. The page gives nothing until the eye receives it. The sentence carries no breath until the reader lends it breath. The meaning does not live only in the ink or only in the mind. It happens in the contact between them.

That is where everything happens.

In the contact.

For this moment to exist, stars had to burn. Hydrogen had to gather under pressure. Carbon had to be forged in stellar interiors and scattered through space by stellar death. A planet had to cool enough to hold oceans. Chemistry had to become biology. Biology had to learn hunger, danger, memory, reproduction, grief. Bodies had to survive. Bloodlines had to continue. Language had to rise from the throat. Symbols had to be carved, printed, typed, remembered.

Then I had to sit down and write.

Then you had to arrive.

The odds are not small. They are absurd. So absurd they stop being numbers and become a kind of music.

And yet here we are.

These words began as fragments. Notes scrawled in the margins of sleepless nights. Sentences that arrived fully formed at three in the morning and refused to wait until something more reasonable, more respectable, more daylit. Ideas that interrupted ordinary Tuesdays with extraordinary clarity, as if someone had briefly lifted a veil and said: Look. Remember. Write this down before it dissolves back into the everything from which it came.

At three in the morning, the body tells the truth differently. The house is quiet. The mind has lost some of its armor. The useful little manager that schedules, edits, compares, and worries about credibility has gone off shift. The room has a hum to it. Electricity in the walls. Blood in the ears. Breath moving without instruction.

Then a sentence lands.

Not built. Received.

There is a physical pressure to it. A sentence like that does not arrive politely. It presses from the inside. It changes the body before it becomes language. The chest tightens. The hand reaches. The phone lights up. The pen scratches. The thought has to be caught before the pattern breaks and returns to the dark.

This is not romantic.

This is signal capture.

Some ideas behave like weather. You do not manufacture them. You notice the pressure change. You feel the wind turn. You see the sky darken over the ridge. Then the thing arrives, with or without your readiness. The only choice is whether you are awake enough to receive it.

So I wrote.

Not because I always understood. Often I did not.

Not because I knew where each fragment belonged. Often I did not.

I wrote because certain ideas carry their own force. They do not ask permission. They insist. They make themselves known the way light makes itself known: by touching what it falls on and revealing shape.

I want to be clear from the start: I am not a physicist. What follows will often touch on the language of physics, but I do so as a student of these ideas, not as an expert. I've explored the writings of quantum mechanics, cosmology, and the mathematics of space and time, letting their wonder shape my understanding, but I hold no formal credentials. What I offer is a quiet respect for these mysteries and a deep attentiveness to what I have encountered.

That distinction matters.

Physics is not decoration. Mathematics is not incense. Science is not a costume to wear when the sentence wants to sound important. These disciplines are hard because reality is hard. They require measurement, correction, discipline, and the humility of being wrong in public.

I respect that.

I respect the people who spend their lives letting reality correct them.

What I carry is different, but not careless. I carry attention. I have stood with ideas long enough for their edges to sharpen. I have read about light and time and matter until the words stopped behaving like distant technical subjects and started touching the body. I have watched equations point toward something I could not fully solve but could still feel as structure. I have felt the same order in a page of cosmology that I have felt in the ocean at dusk.

Not the same information.

The same coherence.

The same sense that beneath the noise, something holds.

I grew up on an island in the middle of the Pacific Ocean, a place where the boundary between the visible and invisible has always been thinner than most people are comfortable admitting. Where I live, the land does not merely sit beneath you. It speaks to you. The wind carries information. The ocean is an intelligence. The light that falls through the clouds above the cliffs is not merely photons making their way from the sun — it is something that fills the chest and quiets the mind and makes the word "beautiful" seem like a child's crayon drawing of a cathedral.

On an island, abstraction has a shorter life.

Weather is not an idea. It arrives on your skin. Wind pushes against the body and changes your balance. Salt gathers on your lips. The ocean does not care what you believe about it. It moves according to pressure, gravity, moon, depth, reef, and distance. It feeds you, holds you, teaches you timing, and then, if you forget its power, corrects you without emotion.

The land is the same.

It does not flatter your opinions. It receives your footsteps and says nothing. It holds memory in stone, root, water, and heat. Stand barefoot long enough and you feel that the ground is not dead matter. It is process slowed to the pace of patience.

That does something to a person.

It makes the body listen before the mind explains.

The indigenous people knew this. Not because they lacked science. Because they lived close enough to consequence to understand relationship. They watched cloud, bird, current, season, and silence. They did not need to separate meaning from matter because the world had not yet been broken into departments.

A wave was a wave.

A sign was a sign.

A body knew when to move.

I am someone who has stood at the edge of a canyon at the moment when the afternoon light strikes the iron-red walls and the shadows fall in patterns that have been falling since before any human stood here to witness them, and felt — with every atom in my body, with a certainty that bypassed all argument — that something was being communicated. That this place, this light, this moment was not random. That the universe was not merely happening. It was speaking.

Not in sentences.

Older than that.

The canyon did not explain itself. It did not offer doctrine. It did not care whether I believed in it. It simply stood there while light moved across its face, and the movement of that light altered everything. Flat stone became depth. Shadow revealed structure. Color changed with angle. The walls received the sun and returned meaning through the body.

My breath slowed.

My jaw softened.

Something in my chest opened without asking my permission.

That is how truth often enters.

Not as argument.

As recognition.

You can call it chemistry, and chemistry is present. You can call it pattern recognition, and pattern recognition is present. You can describe the retina, the optic nerve, the visual cortex, the autonomic nervous system, the release of neurotransmitters, the regulation of breath.

All of that is true.

It is not all of the truth.

A map is not false because it is not the canyon. It is useful. It helps. It points. But it does not hold the heat rising off the rock. It does not carry the silence in the ears. It does not reproduce the strange and total feeling that something vast has leaned close without moving an inch.

This book is my attempt to put language around that feeling. Around the conversation that has been trying to happen between the material and the sacred for as long as there have been minds capable of registering the tension between them.

Not to explain the sacred away.

Not to make science soft.

To bring them back into contact.

Because the split was never in reality. The split was in us.

We took the one world and divided it into rooms. Physics in

one room. Mysticism in another. Poetry down the hall. Biology under fluorescent light. Mathematics on a chalkboard. Love somewhere private, often embarrassed by its own seriousness. Then we wondered why the human being felt fractured.

We handed ourselves a broken mirror and called the pieces disciplines.

But reality never agreed to the separation.

A photon does not stop being physics when it becomes a sunset. A heartbeat does not stop being biology when it accelerates with love. A spiral does not stop being mathematics when it appears in a shell, a storm, a flower, or a galaxy. The categories are ours. The continuity is real.

What you are about to read is my attempt to demonstrate something radical and yet utterly self-evident once you allow yourself to explore it: that love, light, mathematics, and science are not separate conversations happening in separate rooms. They are the same conversation. They have always been the same conversation. We simply, somewhere in the long history of human thought, decided to split the building into disciplines, to lock the doors between the rooms, to insist that the physicist and the mystic and the poet had nothing useful to say to each other.

They had everything useful to say to each other.

They have been saying it for centuries.

We simply stopped listening.

Or worse, we listened only to the parts that protected our preferred identity. The rationalist wanted certainty without reverence. The mystic wanted wonder without discipline. The poet wanted beauty without mechanism. The scientist wanted mechanism without surrender. Each one held a piece and mistook the piece for the whole.

But the universe keeps refusing our camps.

Light behaves as wave and particle. Matter and energy convert. Space and time braid together. Observation participates more deeply than common sense wants to admit. The body runs on electrical gradients. The heart generates rhythm. The brain pulses. Cells communicate. The cosmos expands. The flower spirals. The shell curves. The galaxy turns.

Number keeps wearing the mask of beauty.

Eventually even the cynic has to squint.

Something is going on.

The desire to know — the ache, the reaching, the fundamental impulse to extend from nothingness toward something — is not a human invention. It is not merely a byproduct of evolution or a quirk of neurobiology. It is the name this book gives to the deeper movement from stillness toward expression. Physics describes the mechanisms. This book is reaching for the meaning beneath them. That same desire beats in your chest right now. It is the reason you picked up this book. It is the meaning this book assigns to existence.

To say that plainly is dangerous.

Good.

Some sentences should have teeth.

The desire to know did not begin when humans developed language. Language is one late instrument of an older motion. The sprout bending toward light is reaching. The child asking why is reaching. The scientist leaning over data is reaching. The lover trying to say the thing that language cannot quite hold is reaching. The artist returning to the canvas, the musician chasing the note, the monk entering silence, the mother watching her child sleep — all of it is the same movement wearing different bodies.

From zero toward one.

From silence toward utterance.

From isolation toward relation.

From possibility toward form.

This is what I call love.

Not romance. Not sentiment. Not the emotional weather that comes and goes between people. I mean the deeper movement underneath relation itself. The force that reduces distance. The motion that wants contact. The ache in the whole to know itself through the part.

Love is not soft.

Love is structural.

It is the architecture of relation.

I do not ask you to believe anything written here. Belief is a form of effort, a deliberate act of accepting something the mind has not fully confirmed. What I ask is something prior to belief. I ask you to feel. To notice. To hold these ideas the way you hold a seashell to your ear — not analyzing the architecture of calcium carbonate, not calculating the resonance frequencies of the chamber, but simply listening for the sound that the ocean left behind.

Analysis is a beautiful tool.

Use it.

But do not confuse the tool with the whole body of knowing.

You already know another way. You have used it in grief, when the body understood loss before the mind could form a sentence. You have used it in music, when a chord changed the pressure behind your eyes. You have used it in love, when one look carried more information than a page of explanation. You have used it in prayer, even if you do not call it prayer, when silence became full.

That is not gullibility.

That is contact.

Hold the shell to your ear.

The ocean you hear is not literally trapped inside it. The shell shapes surrounding sound. Its chamber filters noise into a pattern the brain recognizes as sea. The sound is local, but the memory it awakens is ancient.

That is how symbols work.

That is how scripture works.

That is how equations work.

That is how beauty works.

A structure catches the world and returns it in a form the body can recognize.

The ocean is there. It has always been there. You are not separate from it. You carry its sound with you, inside you, because the ocean made you. You are made of the same hydrogen that filled the first stars, the same carbon forged in their dying, the same ancient physics that curved the first fabric of space around the first concentration of mass.

Your body is not outside the cosmos.

Your body is the cosmos made temporary, warm, mobile, and aware.

Your blood carries iron. Your bones carry calcium. Your lungs trade gases with the atmosphere. Your skin receives light from a star. Your nervous system turns electrical pulses into memory, fear, tenderness, hunger, and awe.

You are not a ghost trapped in matter.

You are matter learning to look back.

You are not a visitor to the universe. You are a feature of it. A local expression of its deepest nature. Not an accident wandering through a meaningless machine, but the machine becoming intimate with itself. Not separate from creation, but creation arranged in such a way that it can ask what creation means.

That is a holy thing, whether or not you use the word holy.

That is a staggering thing, whether or not you have remembered how to be staggered.

And that nature, I will spend the rest of this book arguing, is love.

— Kepa

• • •

INTRODUCTION

THE QUESTION THAT LIT THE UNIVERSE

"What is the distance between nothing and everything? One thought. One photon. One breath."

Close your eyes for a moment.

Just one moment.

Not a long meditation. Not a formal practice. Not anything that requires a cushion, a teacher, a mantra, or a special arrangement of your body. Just close your eyes and notice the immediate fact of being here.

Feel the weight of your body.

Not the idea of weight. The actual downward pull. The pressure of your hips against the chair. The contact between your feet and the floor. The way fabric touches skin. The way the surface beneath you receives you without opinion. Your body is not floating through abstraction. It is being held by matter.

Feel that.

There are thousands of tiny contact points between you and the world right now. Skin against cloth. Bone under muscle. Air against the inside of your nostrils. Teeth resting inside the jaw. Tongue touching the roof of the mouth. The body is never separate from the world. It is always being pressed, warmed, cooled, supported, signaled, and shaped by what surrounds it.

This is not philosophy yet.

This is pressure.

Now feel the air entering your lungs.

Do not control it. Do not count it. Do not improve it. Just notice that breathing is already happening. The ribs expand. The diaphragm lowers. Pressure changes inside the chest cavity, and air moves in because pressure differences create motion. Oxygen crosses the thin wet membrane of the lungs and enters the blood. Hemoglobin carries it. Cells receive it. Mitochondria use it to help release stored chemical charge from food.

You are being powered.

Quietly.

Continuously.

Without your permission.

Your body has been doing this since the moment you were born. Even in your deepest sleep, while the part of you that says "I" disappears from conscious control, the lungs keep expanding and contracting. The heart keeps pushing. The blood keeps circulating. The cells keep making their ancient bargain with oxygen: bring us the element, and we will keep the fire small enough to live inside.

That is what metabolism is.

Contained fire.

Life is flame held in water.

Feel the subtle warmth wherever light touches your skin. Even now, even through walls and windows, photons are arriving to meet you. Some began at the sun, crossed roughly ninety-three million miles of space, and arrived here in about eight minutes. A number like that becomes useless if you only think it. Ninety-three million miles is too large for the mind to hold as distance.

So bring it down to skin.

A warm patch on your forearm.

A brightness behind your eyelids.

The sun is touching you from another world.

That is not metaphor.

It is astronomy becoming sensation.

The light leaves the solar surface and travels at the speed the universe set as its maximum speed limit: 299,792,458 meters per second. It crosses emptiness that is not empty, passes through fields and dust and the cold black architecture between worlds, enters the atmosphere, scatters into blue, reflects from walls, bends through glass, and lands on your cheek with the softness of a blessing.

Ninety-three million miles.

Eight minutes.

And here it is.

Warm as welcome.

Why?

That question is the ignition point.

Not how. How is beautiful. How gives us equations, telescopes, particle accelerators, microscopes, anatomy, chemistry, and maps. How tells us the photon traveled. How tells us the lungs exchanged gas. How tells us gravity holds you to the surface of a rotating planet.

But why is the deeper blade.

Why cuts through description and asks what the whole thing is doing here at all.

Why is the question that no equation has ever fully answered. Why is the question every human tradition has circled with trembling hands: scientific, spiritual, artistic, mathematical, devotional. Why is the question carved into temples, hidden in lullabies, embedded in funeral prayers, painted on cave walls, whispered over hospital beds, and built into the silence after music ends.

Why does anything exist rather than nothing?

Sit with the violence of that question.

Not violence as harm.

Violence as force.

It tears open the ordinary surface of the day. It takes the cup, the chair, the hand, the sky, the body, the memory, the grief, the child laughing in another room, the star burning ninety-three million miles away, and asks why any of it should be here instead of nothing at all.

Nothing would have been simpler.

No stars.

No atoms.

No hunger.

No beauty.

No loss.

No one to ask the question.

And yet something happened.

Somewhere before time existed, before space had unfolded its dimensions, before matter had thickened into stars and bones and ocean floors and the particular arrangement of cells that is you reading this sentence, there was stillness.

Pure, absolute, unbroken stillness.

Not the stillness of a quiet room, which is only the absence of noise your ears can register.

Not the stillness of a sleeping body, where the brain continues firing, blood continues moving, cells continue repairing.

Not even the stillness of outer space, which is not truly still at all, but alive with fields, radiation, particles, curvature, and quantum motion.

I mean a stillness so complete it has no surface. No edge. No before. No after. No location. No observer. No breath waiting to be taken. No darkness, because darkness requires the possibility of light. No silence, because silence requires the possibility of sound.

A stillness without comparison.

Mathematicians call it zero.

Mystics call it the void.

Theologians point toward it with names that can only fail beautifully.

Every serious tradition eventually arrives at the same threshold, removes its sandals, and stops pretending language is large enough.

In the beginning, there was nothing.

And then nothing wanted to know itself.

This book names the movement from nothing toward something as a desire to know.

Not because it is decorative.

Because it is structurally necessary.

If nothing had no impulse toward expression, nothing would have remained nothing. No motion. No difference. No light. No first event. No field. No time. No you. No me. No page. No breath.

Something moved.

The stillness did not stay still.

Zero leaned toward one.

That leaning is what I call love.

I want to be exact here, because the word love has been used so carelessly that it often arrives covered in fingerprints. It carries romance, disappointment, family, desire, loyalty, ache, comedy, betrayal, longing, memory, and every song anyone ever wrote when the nervous system became too full to stay quiet.

I mean something older.

I do not mean the love that appears and disappears in human relationships. I do not mean an emotion produced by hormones, though hormones are one of the ways love enters the body. I do not mean sentiment. I do not mean softness. I do not mean the fragile little mood that comes when circumstances flatter the ego.

I mean the primal impulse to reach.

To know.

To extend beyond the sealed boundary of self toward the other, the unknown, the possible, the not-yet-formed.

The name this book gives to the movement from zero toward one.

The ache of perfect completeness for the experience of difference.

Love requires distance.

Not separation as abandonment. Distance as relation. If two things are fused completely, neither can behold the other. The eye cannot see itself without a surface to return the image. A hand cannot touch itself as other without folding. A note cannot become music without the interval between tones.

Relation needs space.

Knowing needs contrast.

Love needs the distance between.

That is why the title matters.

The distance between zero and one is not a simple step. It is the birth of difference itself. It is the first measurable interval. The first opening. The first permission for anything to stand apart long enough to be known.

Love is not a physical force in the scientific sense.

Gravity, electromagnetism, the strong interaction, and the weak interaction describe measurable interactions with formal precision. They govern attraction, charge, atomic structure, decay, fusion, matter, and stars.

This book does not confuse categories.

It names the felt meaning beneath the order.

The reaching.

The binding.

The movement from isolation toward form.

Physics describes what holds things together.

Love names why togetherness feels like truth.

Light was its first language.

When stillness reached toward experience, it reached as light. The first great announcement was not a sentence. It was radiance. A release. A propagation. A movement outward at the fastest speed the universe allows.

Light does not ask permission from darkness.

It arrives.

It reveals surface. It creates shadow. It makes location meaningful. It gives the eye something to receive and the mind something to assemble. Without light, the universe could exist, but it could not be seen. Without light, distance could not announce itself. Form could not appear. Color could not happen. The face of another could not be recognized.

Light is how the hidden becomes visible.

Mathematics was love's grammar.

The rules. The ratios. The relationships. The deep consistency by which the reaching does not collapse into chaos. Mathematics is why stars ignite according to conditions instead of mood. Why circles curve the way they curve. Why spirals repeat across shell, storm, seed, and galaxy. Why the universe can be measured at all.

Mathematics is not cold.

Cold is how we taught it.

Mathematics is structure behaving faithfully.

Science is the modern scripture through which we decode the original text.

Not scripture as unquestioned dogma.

Scripture as disciplined reading.

Science kneels before reality by testing it. It asks the world to answer. It submits its claims to correction. It allows the universe to say no.

A laboratory is also a chapel when the person inside it is honest.

And you, reading this, are a verse in that scripture.

Not only a reader of the story.

A word within it.

Your body is made of elements forged in cosmic events. Your eyes are instruments tuned to a narrow band of electromagnetic radiation. Your brain is a wet electrical architecture turning sensation into meaning. Your grief has chemistry. Your love has rhythm. Your memories have pathways. Your breath participates in atmosphere. Your bones answer gravity.

You are not outside the mystery looking in.

You are the mystery looking through a human aperture.

A hue-man aperture.

Light in flesh.

Color made conscious.

This book is organized as a movement through perception. We begin in stillness and return to stillness, but the return is not a circle that wasted its motion. It is a spiral. You come back to the same word with a different nervous system. You return to silence having heard what silence contains.

The stillness does not change.

You do.

Not because you become someone else.

Because your attention has moved.

Because the body has touched ideas that cannot remain only ideas once they are felt.

In the chapters ahead, we will examine the first four dimensions of creation: light, space, matter, and time. Not as cold abstractions owned by engineers and physicists, but as living conditions that make every experience possible. Light reveals. Space relates. Matter holds. Time unfolds.

These are not only dimensions. They are permissions granted to experience.

We will look at the human body as a crystalline instrument of extraordinary sophistication: a receiver, transmitter, processor, and memory field. Not a meat machine. Not a soul container. A living interface where charge, water, tissue, breath, light, and awareness meet.

We will sit with the ego, not as a villain, but as an archive. A storage system. A hard drive of pain, memory, identity, strategy, defense, pride, longing, and learned survival. The ego is not the enemy. The ego is what happens when experience leaves a mark and the mark calls itself "me."

We will follow the electromagnetic pathways of the chakra system and place ancient maps beside modern nervous system anatomy, not to force them into sameness, but to show that serious attention often finds the same mountain from different trails. The ancient cartographers looked inward with disciplined perception. The modern neuroscientists look through instruments. Both are studying the human engine.

We will investigate places on Earth that feel amplified. Places where the body changes before the mind explains why. Certain mountains. Certain waters. Certain temples. Certain rooms after prayer. Some places hold pattern the way a canyon holds echo. Some places organize attention. You feel it in the skin first.

We will sit with the strange beauty of the simulation hypothesis, not as nihilism, but as wonder. To say reality contains code, pattern, rule, rendering, and participation does not cheapen it. It may make it more sacred. A song is not less moving because it has structure. A film is not meaningless because it is projected. A dream is not empty because the mind generates it.

We will examine manifestation, not as wishful thinking, but as the disciplined alignment of attention, behavior, emotional charge, and probability. A thought does not magically move a mountain. But thought moves the body. The body makes choices. Choices alter environments. Environments change outcomes. Attention is not passive. It is steering.

We will extend awareness through dimensions beyond the four we ordinarily perceive, not to escape this world, but to understand why this world feels larger than its visible surface. The unseen is not automatically unreal. Most of what governs you is unseen: gravity, radio waves, cellular signaling, memory, oxygen exchange, love.

Then we will return to the truth beneath every page of this book:

You are not only a wave.

You are the ocean.

That line is not meant to flatter you.

It is meant to correct you.

The wave has shape, direction, height, motion, and temporary identity. It rises, travels, breaks, and disappears. But at no point is it separate from the ocean. Its individuality is real and temporary. Its source is continuous.

That is you.

A temporary form of something continuous.

A crest of awareness rising for a moment, catching light, moving across time, then returning.

Not lost.

Returned.

Some of what follows will sound scientific. Some will sound mystical. Some will sound like poetry. Some will sound like all three at once, because truth in its natural state does not respect the borders we draw around it.

The universe does not honor our departments.

A photon does not stop being physics when it makes a sunset beautiful.

A heartbeat does not stop being biology when it accelerates with love.

A prayer does not stop being human because the nervous system participates in it.

A number does not stop being sacred because it can be calculated.

The truth is whole.

It has always been whole.

Our categories are the fragmentation.

Not the truth itself.

So begin here.

With the breath entering your body.

With the weight of your bones.

With the warmth of light traveling ninety-three million miles to touch your skin.

With the question that has been waiting underneath every other question.

Why does anything exist rather than nothing?

Let us begin where everything begins.

Let us begin with zero.

CHAPTER ONE

STILLNESS
The Zero Point

"Before the first word, there was silence. Before the first light, there was the love that longed for it."

THE ELOQUENCE OF NOTHING

Sit still for a moment.

Not spiritually. Physically.

Feel the room around you. The low pressure of air against your skin. The faint sounds that remain when nothing important is happening. A refrigerator in another room. A car passing somewhere beyond the wall. Your own breath making a small weather system inside your chest. Even in what you call silence, the world is full of motion.

True stillness is harder to imagine because you have never experienced it.

Not once.

Even in your deepest sleep, the brain is active. The heart contracts. Blood moves. Cells repair themselves. Electrical signals travel through the nervous system. Heat leaves the body. Molecules collide. Time continues.

The stillness I am talking about is older than all of that.

No pulse.

No breath.

No vibration.

No before.

No after.

No witness standing nearby to say, "It is quiet."

There is a number that terrified mathematicians for centuries. A number so paradoxical, so philosophically unsettling that entire civilizations refused to acknowledge its existence, and those civilizations were not foolish. They were standing before an idea with teeth. An idea that genuine intelligence recognizes as dangerous because it can dissolve the foundations of other ideas.

That number is zero.

Zero looks harmless on the page. A small loop. A mouth closed around silence. A symbol so familiar we use it without feeling the force of it. But familiarity has made us numb. Zero is not small. Zero is the most unstable idea in mathematics because it names absence while being present enough to be named.

That is the problem.

How can nothing be something?

How can absence have a symbol?

How can emptiness occupy a position?

Think carefully about what zero represents. Not merely the absence of quantity. Not simply the marker between positive and negative on a number line. Zero is stranger and more profound than either description. Zero is not a blank wall. It is not dead space. It is not a storage room with nothing inside.

Zero is the presence of potential.

It is not empty.

It is full — full of everything that has not yet chosen to become.

Zero is the held breath before the symphony begins. Not silence as failure, but silence as readiness. Every violin still. Every bow lifted. Every musician waiting for the first motion of the conductor's hand. The sound is not there yet, but the room is loaded with it.

That is zero.

Zero is the blank canvas before paint touches it. Not because the canvas already displays every painting, but because it can receive any painting. Its emptiness is not a defect. Its emptiness is its capacity. The white space is not meaningless. It is permission.

Zero is the dark theater before the projector starts.

Zero is the seed before the shell splits.

Zero is the closed eye before light returns.

Zero is the womb before the first cell divides.

Do not soften this.

Zero is not nothing.

Zero is readiness without form.

The ancient Babylonians, who invented positional notation and gave us the first sophisticated mathematics, resisted zero for centuries. They had a placeholder — a pair of diagonal wedges to indicate an empty column — but they could not bring themselves to make it a number. They could use absence as a marker, but not as an entity. They could point to the empty place, but they would not give the empty place its own life.

That hesitation makes sense.

To say nothing is something feels philosophically reckless.

It still does.

The Greeks, despite producing Euclid and Pythagoras and

Archimedes, never fully embraced zero as a number. Their mathematics loved form, proportion, geometry, visible relation. A triangle could be drawn. A circle could be contemplated. A ratio could be measured. But zero was more troubling. Zero did not offer a shape in the world. It did not stand in the sunlight and cast a shadow.

How could nothing have value?

How could absence be present?

How could a void participate in order?

These are not primitive questions.

They are the right questions.

It took Indian mathematicians — particularly Brahmagupta in the seventh century — to formalize zero as a number with its own rules of operation. And when they did, mathematics changed at the root. Not because a useful symbol had been added, but because a deeper permission had been granted.

Once zero enters the number system, reality changes.

Not reality itself.

Our access to it.

Zero makes negative numbers possible in a new way. It makes place value powerful. It makes algebra breathe. It makes calculus imaginable. It makes modern computation possible. Every digital system depends on the most primal distinction: off and on, zero and one. Absence and presence. Silence and signal.

Your phone, your bank account, your music files, your photographs, your messages, your entire modern life moves through architectures built from zero and one.

Nothing and something.

Off and on.

No and yes.

Silence and pulse.

The most advanced machines on Earth are built from the oldest question in existence.

Is there something here?

Or not?

Once you admit zero into the number system, you admit an entire philosophy: that nothingness is a state with properties, that absence has structure, that the void is not empty but pregnant.

This is precisely what modern physics discovered when it looked closely at what we call "empty space."

Empty space is not empty.

That sentence should stop you.

The vacuum is not a dead container waiting for objects to pass through it. It is not a cosmic storage unit swept clean of matter. The quantum vacuum — the lowest energy state of a field — still contains fields. It still contains fluctuations. It still has measurable consequences.

What we once called nothing turned out to be active.

At the smallest scales, the vacuum behaves less like a blank page and more like the surface of water under invisible rain. Particles and antiparticles appear and disappear in ways that are not particles in the ordinary sense, but disturbances in fields, brief exchanges, mathematical necessities with measurable effects. The vacuum is not a stage with nothing on it. The stage itself is alive.

The Casimir effect demonstrates this physically. Bring two uncharged metal plates extremely close together in a vacuum, and they experience an attractive force. Nothing appears to be pushing them. No visible hand. No obvious mechanism. But the restrictions placed on quantum field fluctuations between the plates differ from those outside the plates, creating a pressure difference. The plates move.

Nothing pushed.

And yet the plates moved.

That is zero with teeth.

The Lamb shift in hydrogen, the anomalous magnetic moment of the electron — these are not mystical impressions. They are precision measurements showing that empty space contributes to physical reality. The vacuum leaves fingerprints. It participates.

The void is not blank.

It has behavior.

This is where the ancient and the modern begin to face each other across the table.

Every contemplative tradition on this planet arrived at a version of this understanding through a completely different door. Not with particle accelerators. Not with equations. With disciplined attention turned inward until the ordinary noise of the mind thinned enough for the deeper structure to be noticed.

The Buddhists call it śūnyatā — emptiness that is not mere negation, but the absence of fixed, independent existence. Things exist, but not as isolated objects sealed off from relation. Everything arises dependent on conditions. A cup is clay, heat, hand, use, emptiness, and shape. Its emptiness is what allows it to hold water. Its usefulness depends on what is not there.

The Taoists call it wu — the nameless, the uncarved, the source before distinction. The Tao Te Ching points again and again to the power of what is empty: the hollow of the bowl, the space inside the room, the center of the wheel. The useful part is often the part that is not visibly there.

The Christian mystics speak of the Godhead, the ground, the cloud of unknowing. The source before image, before concept, before even the names we use for God. Not absence as atheism. Absence as the overwhelming fullness before form.

The Hindu sages call it Brahman in its unmanifest state — nirguna Brahman, pure being without attributes. The ocean before wave, before foam, before shore, before anyone stands there to say ocean.

And here, in these notes that became this book, I call it zero.

Not because zero is less sacred than those names.

Because zero is clean.

Zero makes no decorative claim. It does not arrive wearing robes. It does not ask to be worshipped. It points to the mystery and says: here. Before everything. This. The prior condition. The ground floor of being from which all upper floors are built.

Zero is the most honest name we have for the most important thing there is.

Not God as personality.

Not void as absence.

Not space as container.

Zero as the naked fact before fact.

The unlit room.

The closed circuit.

The silent instrument.

The place where everything is possible because nothing has yet been forced into one shape.

THE BIRTH OF DESIRE

Now ask the question that changes everything.

Why did zero not stay zero?

This is the question every physicist and every mystic eventually has to answer, avoid, bow before, or build a life around.

Why?

If absolute stillness were truly sufficient unto itself — if the zero point were perfectly complete in its completeness — then the universe would never have happened. There would be no light, no space, no time, no matter, no chemistry, no biology, no consciousness. No stars burning. No oceans heaving. No bone. No blood. No laughter from another room. No grief. No music. No one reading. No one writing.

Nothingness would have remained nothingness forever.

And even forever is too much.

Forever implies time.

Zero had no time.

No clock could tick there because there was no change to measure. No distance could exist because there were no locations to separate. No event could occur because event requires before and after. The zero point is not a long empty hallway stretching endlessly into darkness. It is not darkness. It is not long. It is not a hallway.

Those are images made by minds that already live in space.

Zero is prior to image.

Prior to distance.

Prior to waiting.

And still, something stirred.

Do not rush past that.

Something stirred.

Something that was complete developed, apparently spontaneously, a motion toward expression. Not need in the human sense. Not lack as damage. Not loneliness as psychological discomfort. Something cleaner. Stranger. More structural.

A wanting.

An ache in what cannot ache.

The quantum vacuum fluctuated. Potential became actual. Stillness became motion. Zero became — not one, not yet, not in a single step — but the beginning of the movement from no-thing toward some-thing.

A pressure change.

A first asymmetry.

A lean.

This stirring is what I call love.

Not because love is a pretty word.

Because no other word comes close to naming the movement from sealed completeness into relation.

Love is the desire to know.

Not love as we have been trained to shrink it: romance, attachment, attraction, possession, emotional reward. Those are later forms. Human forms. Nervous-system forms. Beautiful sometimes. Painful often. Real, but not original.

I mean the deeper impulse underneath all of them.

The movement beyond the boundary of self.

The reach toward the other.

The refusal of sealed isolation.

The ache of undifferentiated oneness to experience itself as something, as anything: here and there, this and that, witness and witnessed, knower and known.

Love requires a margin of separation.

This is not sentiment.

It is geometry.

If two things are fused completely, neither can behold the other. A mirror requires distance from the face. A sound requires space to travel. A bridge requires two sides. A circuit requires difference in potential. A kiss requires two mouths, not one.

Relation needs interval.

Music proves this with brutal clarity. A single tone can be pure, but melody requires difference. Rhythm requires gap. Harmony requires distinct notes held together without collapsing into sameness. If every note sounded at once with no separation, music would not become more complete. It would become noise.

Love works the same way.

Oneness had to open.

Not because oneness was broken.

Because oneness wanted to be known.

The sun gives warmth precisely because it is not where you are. Its fire must travel across space before it can touch your skin as light. If the sun and your skin occupied the same point, there would be no warmth as experience. Only fusion. No relation. No receiving. No distance crossed.

Distance makes contact meaningful.

This is one of the great hidden laws.

Too much distance, and relation weakens. No contact. No signal. No warmth.

No distance, and relation disappears into merger.

Love lives in the charged interval between.

The universe, to know itself, could not remain a perfect, undivided whole. It had to open a space within itself. It had to become both witness and witnessed, seeker and seen, knower and known.

In that first distance, consciousness became possible.

In that first distance, love became physical.

The act of creation was the act of the universe making itself into two — not out of failure, not out of deficiency, not because something was wrong with the original unity, but because experience requires relation.

God — consciousness — the source — I use these words with respect, knowing each is both useful and insufficient — did not create the universe out of boredom or loneliness or some cosmic therapeutic need.

It created out of desire.

Out of *love*—the desire to know:

What is it like?

Where is it?

How does it feel?

When does it change?

Who is there?

Why is anything here at all?

These six questions — what, where, how, when, who, and why — are not merely human inventions. They are older than grammar. Older than biology. Older than the mouth that speaks them.

They are the deep coordinates of existence.

What asks for identity.

Where asks for space.

How asks for mechanism.

When asks for time.

Who asks for consciousness.

Why asks for meaning.

Together, they form the first operating system of reality. The primal inquiries embedded in existence from the beginning. The first permissions. The first code running in the first instant of becoming.

We ask them because we are made of the same movement that first asked them.

Your curiosity is not separate from cosmic curiosity.

It is cosmic curiosity wearing your face.

When a child asks why the sky is blue, the universe is not entertaining itself with a cute developmental stage. It is continuing the same ancient motion that produced light, atmosphere, scattering, eyes, language, and the soft human need to understand.

Curiosity is love in question form.

It reaches.

It crosses the distance between not knowing and knowing.

It opens the closed room.

This is why deadening curiosity is one of the great violences of culture. When a child asks why and receives annoyance instead of attention, something sacred is being mishandled. Not because every question must be answered perfectly, but because every question is an expression of the original reach.

The desire to know is not childish.

It is cosmic.

THE MATHEMATICS OF LONGING

Consider the number line.

Do not imagine it as a clean diagram in a textbook. Feel it as a physical structure. A wire pulled tight across the dark. A road extending beyond sight in both directions. At one point, near the center of your attention, zero.

Extending to the right: one, two, three, a thousand, a trillion, numbers so large language begins to fail.

Extending to the left: negative one, negative two, debt, absence, inversion, below, before, loss, antimatter in the imagination of arithmetic.

Then between the marks: fractions, decimals, irrational numbers, values that refuse to resolve neatly.

A whole universe of position.

Now consider only the interval between zero and one.

Just that tiny span.

The distance your mind wants to treat as almost nothing.

What lives there?

Infinity.

Not poetic infinity.

Mathematical infinity.

0.1.

0.01.

0.001.

0.0001.

Keep going. The decimals continue without exhaustion. Between zero and one are halves, thirds, sevenths, square roots, repeating decimals, non-repeating decimals, numbers that can be named, numbers that cannot be named, numbers that exist whether or not any mouth ever speaks them.

Between any two numbers, no matter how close together you imagine them, there is another number.

And another.

And another.

Always.

This should disturb you.

The line between zero and one is not empty space. It is infinitely dense. It contains more than the mind can count, because counting itself is too crude an instrument for this kind of fullness.

Georg Cantor proved in the 1870s that the infinity of real numbers between zero and one is larger than the infinity of counting numbers. That statement sounds impossible at first because the mind wants infinity to be one thing: endlessness. But Cantor showed that infinities have size, that the endless can have deeper orders of endlessness, that the continuum between zero and one cannot be placed in one-to-one correspondence with the integers.

There are more points between zero and one than there are counting numbers stretching without end.

Sit with that until it becomes physical.

The small gap is not small.

The space you thought you could step over contains an abyss.

The distance between zero and one holds a deeper mystery than the distance between one and a trillion. The passage from one to a trillion is accumulation. More of what already exists. Bigger piles. Larger distances. Greater quantities.

But the passage from zero to one is different.

It is not increase.

It is emergence.

It is the birth of being.

Everything after one is arithmetic.

Zero to one is creation.

A machine understands this. A computer does not need ten thousand mystical symbols to build an empire of complexity. It needs zero and one. Off and on. No charge and charge. Closed gate and open gate. From that binary pulse, worlds are rendered: images, voices, maps, markets, memories, films, songs, simulations, entire architectures of modern life.

The complexity comes later.

The first miracle is the switch.

No signal.

Signal.

Zero.

One.

A house works the same way. Before the first board is placed, the structure exists as potential. Plans on paper. Measurements. Load paths. Materials stacked nearby. But the house is not yet a house. Then the first stake is driven into the ground. A line is established. A corner is marked. Orientation appears. From that first point, every other measurement becomes possible.

Zero to one.

The same with a seed. Before the shell splits, the tree exists as encoded potential. Not imaginary. Real, but unexpressed. Root, trunk, branch, leaf, shade, fruit, all held in compact instruction. Then moisture enters. Pressure builds. The shell breaks. A root tip pushes downward. A shoot reaches upward.

Zero did not become forest in one leap.

It became one rupture.

One direction.

One reach.

That is how creation moves.

Not all at once.

Through first permission.

This is not merely mathematical curiosity. This is a description of reality.

The distance between zero and one — between nothing and the first expression of something — is not a gap to be leaped across. It is an infinity to be explored.

Every decimal place is a dimension of experience.

Every fraction is a possible angle of being.

The universe did not leap from nothing to something and finish the work. It is still crossing that distance. It is still discovering new values between absence and fullness. New forms. New minds. New arrangements of matter. New ways for the whole to encounter itself as part.

You are one of those points.

A particular value between zero and one.

A specific coordinate in the infinite density of possibility.

No other point is quite where you are.

This is not motivational language.

It is mathematics turned inward.

When Cantor presented his mathematics, the establishment of the nineteenth century reacted with hostility. Leopold Kronecker, one of the eminent mathematicians of the era, famously attacked Cantor and his ideas. The resistance was not merely technical. It was emotional. The idea that infinity contains infinities, that some infinities are larger than others, that the territory between two apparently adjacent points is bottomless — this shook something deeper than notation.

People do not only resist falsehood.

They resist truths that break the container they built their identity inside.

Infinity does that.

So does stillness.

So does love.

They all reveal the same humiliation: what we thought was simple was not simple. What we thought was empty was not empty. What we thought we understood was only the top layer of something without a visible bottom.

The mathematics of infinity disturbs us for the same reason silence disturbs us. Sit quietly long enough and you discover that silence is not empty. It contains everything you have been avoiding. Memory rises. Fear rises. Desire rises. Old grief knocks from under the floorboards. The room was never empty. It was only unoccupied by noise.

Zero is like that.

It waits.

Then it reveals what was held in potential.

The inner movement is vaster than the outer one. The distance between silence and the first sound contains more than the distance between any two stars.

Because stars already exist.

The first sound has to cross the impossible.

THE PRACTICE OF ZERO

Find a quiet place.

Sit comfortably.

Close your eyes.

Let the body make contact with the surface beneath it. Feel the weight transfer downward. Let the chair hold you. Let the floor hold the chair. Let the Earth hold the floor. Gravity is already doing its work. You do not have to believe in it. You are being held by it.

For three minutes, do nothing.

Not meditation with a technique.

Not mindfulness with a method.

Not a breathing exercise.

Not a visualization.

Not a mantra.

Nothing.

Let the breath happen without instruction. Let the jaw release.

Let the tongue rest. Let the shoulders drop by whatever amount they are willing to drop. Let the hands become heavy.

Thoughts will come.

Of course they will.

The mind is not a mistake. It is an instrument built to generate, compare, predict, remember, rehearse, protect, and interpret. Asking it to produce no thoughts is like asking the ocean to stop making waves. The point is not to freeze the water. The point is to notice the depth beneath the surface motion.

When a thought appears, do not chase it.

Do not fight it.

Do not improve it.

Notice its arrival.

It comes from silence, takes shape, occupies attention briefly, then changes or disappears. Like a virtual particle arising from the vacuum, it appears from the field, participates for a moment, then returns.

The silence that receives it is not damaged.

The silence is revealed by the arising and the return.

You cannot see the ground while you are running.

Stillness lets the ground appear.

Notice what emerges from the nothing.

A sound you had not registered: the hum of a device, a bird outside, the slight movement of air through the room.

A sensation in the body: warmth in the hands, pressure behind the eyes, a small ache in the lower back, tingling in the foot, the pulse in the neck.

A memory.

An emotion waiting for a gap in the noise.

An image, rising like a whale from deep water.

Do not make a project out of it.

Do not turn every arising into homework.

Not everything that appears needs interpretation. Some things only need to be noticed and allowed to return.

This is a crucial distinction.

The ego wants to grab whatever arises and turn it into identity, problem, plan, or proof. Zero asks for something else. It asks you to witness without seizing. To let form emerge and dissolve without immediately making it useful.

This is hard because modern life has trained the nervous system to fear open space. Silence feels inefficient. Stillness feels suspicious. Doing nothing feels like falling behind.

That is sickness disguised as productivity.

The soil does not apologize for winter.

The tide does not explain its retreat.

The seed does not panic in darkness.

Zero is not wasted time.

Zero is the condition that allows form to return clean.

This is your zero point.

You can return here anytime.

You have never actually left it.

Behind the thought, stillness.

Behind the breath, stillness.

Behind the pulse, stillness.

Behind the ache, the plan, the worry, the identity, the name, the memory, the ambition — stillness.

Not far away.

Underneath.

The practice takes three minutes. Do it daily for thirty days and observe what changes in the quality of your attention. The texture of your awareness. The way ordinary moments begin to feel charged with something that was always there, quietly waiting to be noticed.

You may find that the pause before speaking becomes wider.

That anger still rises, but you see it sooner.

That beauty lands harder.

That grief moves with less resistance.

That your body has been whispering for years and you were simply too loud to hear it.

This is not escape.

This is return.

Zero is not the absence of life.

Zero is where life gathers before it speaks.

CHAPTER TWO

LIGHT
Love Made Visible

*"Light is the first word love ever spoke.
It is still speaking."*

THE FIRST ACT

Feel light on your skin.

Do not think about the sun yet. Do not turn it into astronomy. Start with the body. Start with the warmth across the back of your hand. The slight narrowing of your eyes when brightness enters the room. The way your pupils contract without asking permission. The way your skin can tell the difference between shade and sunlight before your mind makes a sentence about it.

Light arrives first as contact.

Then as vision.

Then as meaning.

In the beginning — and I use those words knowing they are paradoxical, because "beginning" implies time and time did not yet exist, because "in" implies space and space had not yet unfolded — stillness desired to know itself. And the mechanism of that knowing was light.

Not language.

Not thought.

Not matter.

Light.

The first announcement was radiance.

The first movement from hidden to revealed.

The first act of the invisible becoming visible.

Consider what a photon actually does. Consider what it is for a photon to exist.

A photon is created when an atom or molecule releases electromagnetic radiation — when an electron drops from a higher energy state to a lower one, when a charged particle accelerates, when matter changes condition and emits a packet of light. A system shifts, and that shift does not vanish. It travels. It departs as information.

This is already profound.

A change happens somewhere, and light carries the news.

From the moment of its creation, a photon extends outward, always outward, always reaching. It does not pause. It does not reconsider. In a vacuum, it moves at 299,792,458 meters per second — the maximum speed at which information can travel through the universe.

Light is not lazy.

Light does not negotiate with distance.

It crosses.

If nothing absorbs it, scatters it, bends it, or blocks it, it continues. Across rooms. Across oceans. Across interstellar dark. Across billions of years. There are photons traveling through the universe right now that began their movement before Earth had trees, before eyes existed to receive them, before any creature had ever lifted its face toward the sky.

They are still moving.

That is commitment.

Think about what this reaching means. The photon does not illuminate what it prefers. It illuminates what it encounters. It arrives at the ocean and the stone, the window and the wound, the eye of a hawk and the grain of sand on a beach no human foot will touch. It does not choose its recipients. It does not hold itself back for the worthy.

Light gives itself to contact.

This is not soft.

This is exact.

When light strikes matter, something happens. It reflects, refracts, scatters, heats, excites electrons, initiates chemical reactions, triggers photosynthesis, allows vision, sets circadian clocks, burns skin, grows forests, powers weather, reveals distance, creates shadow.

Light does not merely show the world.

Light participates in making the world active.

This is not incidentally like love. This is structurally love.

The photon is one of the universe's purest expressions of its fundamental impulse: to reach, to touch, to reveal, to give itself entirely without becoming less itself. One candle can light another without losing its flame. One sunrise can touch every window in a valley without dividing its intention. One beam can enter a room and make every hidden surface confess its shape.

Light is love given velocity.

Love is light experienced as meaning.

Do not make that vague.

A flashlight in a dark room proves it. Before the beam, the room exists but remains unavailable to perception. The chair is there. The wall is there. The door is there. But without light, relation is impaired. Distance becomes uncertain. Edges disappear. The body moves cautiously because it cannot read the space.

Then light enters.

Instantly, the world becomes navigable.

That is what love does.

It makes relation visible.

A lighthouse proves it another way. The light does not drag the boat to shore. It does not calm the storm. It does not remove the reef. It simply reveals orientation. It gives the vessel a point of relation in the dark. The sailor still has to steer. The ocean still has force. But now there is bearing.

That is love too.

Not rescue.

Orientation.

A fiber optic cable proves it mechanically. Information encoded in pulses of light travels through glass across great distances. Voice, image, memory, money, map, message — all carried by light constrained inside a channel. The content changes. The carrier remains light.

This is what the body is doing with meaning.

This is what a book is doing with your eyes.

This is what the sun is doing with the Earth.

Now consider the physics of light's relationship with time.

Albert Einstein's special relativity demonstrated that time and motion are not separate in the way ordinary experience suggests. As an object with mass approaches the speed of light, time relative to that moving object slows in relation to an outside observer. The mathematics become extreme near light speed. At light speed itself, the ordinary notion of time breaks open.

A photon does not carry a wristwatch.

A photon has no rest frame in the way an object with mass does. We cannot properly speak of "the photon's experience" as if it were a tiny traveler looking out a window. But the equations point toward something extraordinary: along the path of light, the separation between emission and absorption collapses in a way that refuses our ordinary sense of duration.

For us, sunlight takes about eight minutes to travel from the sun to Earth.

For light, the story is stranger.

The beginning and arrival are linked by the geometry of spacetime in ways that challenge our ordinary sense of duration.

Light moves at the boundary where our usual ideas of time begin to break down.

That is enough.

Mystics have described this state for millennia: the eternal present, the living now, the timeless ground. Not because they were doing relativity with prayer beads, but because disciplined attention can reveal something that physics later names in another language: the present is not a thin slice between past and future. It is the only place anything ever touches reality.

Your body knows this.

You cannot breathe yesterday.

You cannot feel tomorrow's sunlight.

You can remember. You can anticipate. But contact only happens now.

Light is the great teacher of now.

It arrives only here.

HUE-MAN: BEINGS MADE OF LIGHT

The word "human" carries a hidden poetic truth: hue-man.

A being of hue.

A being of color.

A being whose entire visual world is defined by the frequencies of electromagnetic radiation reflected, absorbed, emitted, and perceived.

Strict linguistics can keep its records. Let it. The documented origin of human is one conversation. This is another. Hue-man is not etymology. It is recognition.

You are a creature of light.

You do not merely see color. You are shaped by it. Red raises alertness. Blue morning light wakes the body. Darkness releases melatonin. Firelight softens the room and the nervous system with it. Green landscapes change the breath. The gold hour before sunset makes the world feel briefly forgiven.

Color is not decoration.

Color is frequency entering biology.

Your skin responds to light. Your eyes respond to light. Your hormones respond to light. Your sleep responds to light. Your mood responds to light. Your immune system, metabolism, and daily rhythm all participate in the conversation.

This is not fantasy.

This is physiology.

Every day, sunlight enters the human body and begins work. Ultraviolet B radiation initiates vitamin D production in the skin. Morning light enters the eyes and helps regulate the circadian clock through specialized retinal cells sensitive to light. Brightness tells the brain when to wake, when to release hormones, when to prepare digestion, when to sharpen attention, when to begin the long descent toward sleep.

Light is not outside you.

Light is instruction.

Your body reads it.

A plant makes this obvious. Place it near a window and it leans. Chlorophyll receives light and uses it to drive photosynthesis, converting sunlight, carbon dioxide, and water into sugars that feed the living structure. A tree is organized sunlight standing upright in wood.

You are not a plant.

But you are not separate from that process either.

You eat the light plants captured. You eat the animals that ate the plants. The calories in your body are sunlight translated into chemistry. When you walk, think, speak, love, build, grieve, and pray, you are spending stored sun.

You are matter organized by light, powered by stored sunlight, and tuned to receive light as warmth, vision, rhythm, and meaning.

Even your cells emit faint light. Living systems produce ultra-weak photon emissions associated with metabolic reactions and oxidative processes. Sensitive instruments can detect this faint biological glow. The body is not visibly shining like a lamp, but at the cellular level, light is part of the story.

This matters.

Not because it proves every mystical claim anyone has ever made about auras or radiance. Because it grounds the intuition that life and light are intimate.

The body speaks electrically.

The body signals chemically.

The body also glimmers.

Quietly.

Below the threshold of ordinary sight.

That is enough to change how you understand yourself.

You are not a dark object being illuminated from outside.

You are a living system in constant exchange with light.

The deeper implication is simple and enormous: if light was the first language of creation, and if your body depends on light, responds to light, organizes itself by light, and emits light at the cellular level, then your relationship to the source is not external.

It is not the relationship between a lamp and a stranger standing nearby.

It is the relationship between flame and ember.

Between ocean and wave.

Between sun and warmth on skin.

Shared process.

Shared substance.

Spiritual traditions across cultures often speak of us as reflections of the divine. Read that through light and it becomes physical. A source expressing its nature through form. Radiance becoming biology. Order becoming tissue. Awareness becoming face.

We are made in the image of light.

Not because a story says so.

Because the body says so.

Consider the partnership between sunlight and your body. When sunlight contacts your skin, it does not merely warm the surface. It initiates reaction. When morning light reaches your eyes, it does not merely help you see. It calibrates clocks deep in the brain. When darkness comes, the absence of light becomes instruction too. The body releases different signals. Muscles soften. Temperature shifts. The mind moves toward dream.

Light and darkness are not opposites in a moral drama.

They are paired instructions.

Day says rise.

Night says return.

The body needs both.

A camera understands this. Too much light and the image blows out. Too little and the image disappears into noise. A good exposure requires relationship: aperture, shutter, sensitivity, source. Seeing depends on balance.

The nervous system is the same.

Too much stimulation burns the signal.

Too little and the system dulls.

Life is exposure management.

The spiritual traditions knew this symbolically. Light as revelation. Darkness as mystery. Dawn as renewal. Fire as presence. Halo as sanctity. Candle as prayer. Sun as source. These symbols endured because they are not arbitrary. They are anchored in the body's direct experience.

When light enters darkness, the body understands.

When a candle is lit in a room of grief, something changes.

Not because wax and flame solve death.

Because light gives the nervous system a point to gather around.

A small, visible defiance of total dark.

That is why we light candles.

That is why temples glow.

That is why stained glass matters.

That is why sunrise still works on people who claim they believe in nothing.

Light entering your body is not a resource you consume. It is a conversation you are having with the sun.

And the sun, whatever else it is, is a fusion reactor converting hydrogen into helium, releasing vast amounts of energy as light and heat. Matter becomes radiance. Mass becomes message. Pressure becomes morning.

That message crosses ninety-three million miles and lands on the forearm of a being whose body is also built from star-formed elements, whose cells spend stored sunlight, whose eyes evolved to receive this narrow band of electromagnetic information and call it world.

That is intimacy.

LOVE IS LIGHT INTERNALIZED

If light is love made visible — desire given velocity and direction — then what is the experience of love inside the human body?

Start there.

In the body.

Think of someone you love without defense. Someone whose face changes your breathing. Someone whose presence lowers the weaponry in your chest. Notice what happens before the mind explains it.

Warmth.

Softening.

Expansion.

The mouth changes. The shoulders drop. The eyes alter their focus. The heartbeat may quicken or steady depending on the kind of love. Protective love brings heat and readiness. Tender love brings release. Grief-love brings pressure behind the eyes and weight in the sternum.

Love is not only an idea.

It is physiology.

Neurochemistry participates: oxytocin, dopamine, serotonin, endorphins, stress hormones rising or falling according to safety, attachment, longing, bonding. The nervous system changes state in the presence of love. The body reconfigures around relation.

But chemistry is not the whole story.

The deeper pattern is coherence.

When you are loved well, the body becomes more organized. Breathing steadies. Muscles release. Attention becomes less scattered. The system stops scanning for threat and begins to receive the present moment. Love brings the organism into more usable order.

That is what real love does.

It does not make you smaller.

It makes you coherent.

The heart is central here, not as a greeting card symbol, but as an electrical organ. Every heartbeat is an electrical event before it is a muscular one. The sinoatrial node fires. Electrical activity spreads through cardiac tissue. Muscle contracts. Blood moves. The rhythm of the heart can be measured, tracked, interpreted.

The heart generates an electromagnetic field.

The brain does too.

The body is not a bag of chemicals. It is an electrical ecology.

When people speak of feeling someone's presence before a word is said, the statement is not ridiculous. Bodies read bodies constantly. Posture, breath, facial micro-movement, pupil size, vocal tone, skin temperature, rhythm. We are transmitting and receiving more than sentences. We are broadcasting state.

A calm person changes a room.

An angry person changes a room.

A grieving person changes a room.

A loving person changes a room.

You know this because you have felt it.

When you love someone, your body organizes around their reality. Their pain registers in you. Their joy registers in you. Their absence has physical weight. Their voice can alter your breath through a phone speaker thousands of miles away.

This is not weakness.

This is relation doing its work.

Love is not merely a feeling privately owned by one nervous system. Love is a field event between bodies, memories, signals, rhythms, and meaning. It is biology, electricity, attention, and story entering coherence around another being.

The warmth of sunshine and the warmth of being loved feel similar because both enter the body as safety, illumination, and contact. One arrives through photons. One arrives through presence. Both tell the nervous system: open.

The body does not lie about their kinship.

It knows.

This is why a child reaches for a parent in fear. Not for an explanation. For regulation. For warmth. For the field of another body organized enough to help their own body return from alarm. The parent's chest, voice, hands, breath — these are instruments of light in human form.

A hug is architecture.

Pressure around the ribs. Heat transfer. Breath synchronization.

Heartbeats in proximity. The body receives the message through skin before language can say, "You are safe."

That is love made physical.

That is light internalized.

LIVING INTEGRATION: BECOMING THE LIGHT

Step outside.

If it is daytime, find a patch of direct sunlight. Stand there without performing anything. Let the light touch your face, arms, hands, whatever skin is available. Feel the exact location where warmth begins. Notice the border between sun and shade. One inch of difference changes sensation. That is how precise light is.

If it is evening, find the last glow on the horizon. Watch how the world changes when the sun lowers. Edges soften. Colors deepen. Shadows lengthen. The same objects become emotionally different because the light has changed.

If it is night, find a star. Any star. Hold your face toward it and understand that what reaches your eye is ancient travel arriving as present contact.

Light is always crossing distance.

As the light touches you, do not treat it as a resource delivered from elsewhere. Treat it as a relative recognizing you across distance. This light and the light inside your cells belong to the same universe. They are not strangers. They are expressions of one system in different states: radiance, chemistry, tissue, sensation, awareness.

Now close your eyes.

Sense the light inside you.

Not metaphorically.

Physically.

Feel the warmth in your chest, slightly left of center, where the heart contracts and releases. Feel the electrical current in your nerves as tingling, aliveness, readiness. Feel the luminous field behind your closed eyelids, the shifting patterns the visual system produces even in darkness.

Your body is not dark inside.

It is active.

Charged.

Signaling.

Alive with small lightning.

Sit with this for five minutes. Do not force transcendence. Do not chase vision. Let the body register the obvious miracle it usually ignores: you are a living receiver of starlight, walking around as matter that learned to feel.

Then open your eyes.

Look at the room.

Notice what light is doing.

It is revealing every surface.

It is giving you the world.

You are not walking through light.

You are light, walking.

CHAPTER THREE

THE FOUR DIMENSIONS OF CREATION

"Light weaves the fabric of space, allowing matter to dance through the infinite unfolding of time. This is the architecture of everything."

THE BLUEPRINT

Hold up your hand.

Turn it slowly in the light.

Watch what happens.

The surface changes as the angle changes. Knuckles brighten. Lines deepen. Shadows appear between fingers. The hand is the same hand, but the light reveals different truths depending on where it touches. Move the hand through space and the world behind it shifts. Close the fingers and matter presses against matter. Wait one second, then another, and the gesture becomes sequence.

Light.

Space.

Matter.

Time.

You are watching the architecture of reality operate at the end of your arm.

This is not theory.

This is the room you are inside.

We live inside a four-dimensional masterpiece. This is not a figure of speech. It is a precise description of the structure that houses every experience you have ever had or will ever have. It makes the cup of coffee possible. The conversation possible. The grief possible. The joy possible. The particular quality of light on a winter afternoon possible. The smell of rain on pavement. The weight of a sleeping child against your chest. The sound of someone's voice after a long absence.

All of it needs architecture.

Experience does not float in nowhere.

It requires conditions.

Four dimensions. No more, no fewer, for the human field of ordinary perception. A room perfectly sized for the kind of experience it was designed to contain.

Most of us learned dimensions as dry geometry in school: length, width, depth, and time. A coordinate system. A graph. A grid for locating objects and calculating trajectories. Useful for engineers, navigators, architects, physicists, pilots, builders, surveyors.

Useful, yes.

But dead on arrival as a description of what it feels like to exist.

A grid does not explain the ache of distance.

A coordinate does not explain why a childhood bedroom can still live in the body decades after you left it.

A diagram does not explain why time moves differently in grief than it does in joy.

The technical description is not wrong.

It is incomplete.

What if the dimensions are not merely a grid?

What if they are not only a system for locating objects, but a set of permissions that allow experience to occur?

What if each dimension is not a direction in which things extend, but a specific type of relationship that love requires in order to know itself?

That changes everything.

Because then the universe is not just a container.

It is an instrument.

And the dimensions are the strings.

Let me tell you what I see when I close my eyes and feel the architecture of reality rather than only think it.

The first dimension is light.

Not length.

Light.

The initial reach. The desire made manifest. A single ray extending from source toward destination, from the zero point of stillness toward the first coordinate of experience. A line of pure intention moving outward.

A flashlight in a dark room teaches this better than a textbook. Before the beam, there is uncertainty. You know the room exists, but you cannot relate to it clearly. The beam extends. Suddenly direction appears. Forward. There. That wall. That chair. That door. Light creates the first usable relationship between the witness and the world.

The first dimension is the line of contact.

The arrow.

The reach.

Before there was width, before there was depth, before there was duration, there was this: one ray of love moving outward from the place where nothing was, reaching toward the place where something could be.

The first dimension is the universe's first verb.

Reach.

Not reach as desperation.

Reach as emergence.

Reach as the first refusal of sealed stillness.

A laser shows this mechanically. It does not flood the whole room. It concentrates. It moves in one direction with coherence. It carries power because its waves are aligned. A scattered light bulb illuminates broadly. A laser cuts steel because intention has been focused into a line.

That is the first dimension.

Focused desire.

A direction before a world.

The second dimension is space.

The ray discovers that it can turn. That there is not only this direction, but also that direction. One line meets another and creates angle. Angle creates relation. Relation creates geometry.

Now there is here and there.

This point and that point.

Distance.

Difference.

A plane.

Space is what allows the first reach to become relationship instead of mere extension. Space opens the field. It gives the ray room to encounter another ray. It creates surface, boundary, nearness, farness, left, right, above, below.

A blueprint shows this. A single line can mark an edge, but a room does not appear until lines meet. Corners create enclosure. Walls create inside and outside. The house begins as relationship on a plane: measurements, angles, openings, paths. Before lumber is lifted, before concrete hardens, before the roof casts shade, space has already been organized.

Space is the field of relation.

It is love making itself large enough to contain difference.

You cannot love in the absence of space. Love requires a lover and beloved, witness and witnessed, signal and receiver, the face and the mirror, the voice and the ear, the hand and the hand it reaches for.

Space is not emptiness.

Space is the distance that makes relation possible.

Too much space, and the signal weakens.

No space, and relation collapses into sameness.

Love needs enough distance to cross.

The third dimension is matter.

The plane deepens. Flatness acquires volume. Surface becomes body. Geometry becomes object. Two dimensions of relationship give birth to the possibility of substance: something that can be held, dropped, lifted, broken, built, buried, worn, eaten, carved, kissed, carried.

Matter is the point at which the universe pushes back.

Touch the table.

Feel the resistance.

Your hand does not pass through because electromagnetic forces between atoms resist compression. What you experience as solidity is not dead stuff. It is structured interaction. Fields repelling. Particles arranged. Forces maintaining form.

Matter is agreement held under tension.

A wall stands because atoms, bonds, gravity, load, and design are cooperating.

A bone holds because mineral and collagen are organized into strength.

A cup holds water because clay was shaped, fired, hardened, and made into boundary.

Matter is light made patient.

Einstein gave us the conversion equation — $E=mc^2$ — and we teach it as if it belongs only to nuclear reactions and chalkboards. But beneath the equation is a staggering statement: mass and energy are not separate substances. They are related expressions. Matter is concentrated energy. Energy can become mass. The universe is not divided between the luminous and the solid as cleanly as the senses suggest.

Matter is the universe becoming touchable.

Love made solid enough to hold.

A house is love in the form of walls if it shelters the body.

A bowl is love in the form of emptiness held by matter if it feeds someone.

A hand is love in the form of flesh if it reaches.

Matter is not the opposite of spirit.

Matter is spirit accepting weight.

The fourth dimension is time.

Once light has reached, space has opened, and matter has taken form, something essential is still missing: change.

Without time, every event would be collapsed into a single unsorted totality. Every note of every song sounding at once. Every face you have ever loved appearing simultaneously. Birth and death without sequence. Seed and tree and fallen leaf occupying the same undifferentiated instant.

No development.

No anticipation.

No memory.

No becoming.

Time gives experience order.

It lets the universe unfold instead of explode into total simultaneity.

Time is the dimension of sequence. The mechanism by which the whole can be tasted one moment at a time. It is why music works. Why a story works. Why healing works. Why grief has waves. Why a child grows. Why dough rises. Why wounds close. Why dawn matters after a long night.

Time is love's patience with itself.

The universe did not express everything at once because everything at once cannot be experienced as meaning. Meaning requires sequence. First this. Then that. Tension. Delay. Arrival. Loss. Return.

A song proves this.

Take every note of a symphony and play them at the same instant. The information is technically present, but the music is destroyed. Order requires time. Beauty requires timing. The rest between notes is not absence. It is structure.

Time is the rest that lets the note matter.

We treat time like an enemy because it takes everything from us. But time also gives everything to us. Without time, nothing could arrive. Nothing could ripen. Nothing could be forgiven. Nothing could be remembered. Nothing could become.

Time wounds.

Time also opens.

That is the bargain.

THE THREAD OF LIGHT SPEED

Here is the detail that stops me every time I encounter it: the thread connecting all four dimensions is the speed of light.

Not as trivia.

As structure.

In Einstein's formulation of spacetime, the speed of light is not merely the velocity that light happens to travel. It is the invariant limit, the conversion factor, the deep ratio linking space and time. It tells the universe how distance and duration relate. It holds causality together.

One second of time corresponds to about 300,000 kilometers of light-travel distance.

That is not just a big number.

That is the loom.

Space and time are not two separate containers sitting beside each other. They are woven. The speed of light is the weaving tension. Pull one thread and the other moves.

This means light — love made visible — is not merely the content of the first dimension. It is the thread running through the entire structure. Light reveals space. Light interacts with matter. Light sets the limit by which time and causality maintain order.

Without that limit, the universe becomes incoherent.

Cause and effect would lose sequence. Signal would arrive before event. Information would have no discipline. Reality would become a room where every conversation happens at once and no voice can be understood.

The speed of light is not just speed.

It is cosmic manners.

It keeps the universe from interrupting itself.

Light creates space by making distance visible. A room in darkness may still have dimensions, but light makes those dimensions available to perception. It tells the eye where the wall stands, where the floor drops, where the hand ends and the object begins.

Light reveals the field.

Light creates matter not by casually slowing down like a car entering traffic, but by participating in the deep equivalence between energy and mass, by being part of the processes through which the early universe cooled, particles formed, atoms stabilized, stars ignited, chemistry unfolded, and eventually bodies emerged. The world is not frozen light in a simplistic sense. It is energy organized into persistent form.

Still, the poetic truth stands because the physics allows it:

Matter is radiance under discipline.

Light creates time by establishing the maximum rate at which information travels. This limit allows before and after to mean something. It keeps the universe readable. It allows the story to unfold in order.

A network engineer understands this. Signal speed matters. Latency matters. Propagation delay matters. If packets arrive out of order, systems fail or must correct. Information needs timing. A universe is the ultimate network, and light speed is the protocol limit.

There it is.

The mystic and the engineer shaking hands in the server room.

Light is the thread.

Love is the movement in the thread.

The four dimensions are the pattern.

And we are woven into them so completely that there is no clean place where we end and the fabric begins.

Your body proves it right now. Light enters your eyes. Space holds your posture. Matter gives you weight. Time allows this sentence to become one word after another instead of a meaningless pile of language.

You are not inside the universe like a marble inside a jar.

You are the universe locally folded into perception.

ART AS DIMENSIONAL MIRROR

I want to sit with the question of why human beings make art.

Not the shallow answer.

Not because we are bored.

Not because we need decoration.

Not only because art signals status, attracts mates, builds community, preserves memory, or communicates emotion — though all of that has truth in it.

The deeper answer is this:

Art lets the universe rehearse its own structure in miniature.

A painting is light captured in matter, held in space, viewed across time.

Pigments absorb certain wavelengths and reflect others. That is light. Canvas, wood, linen, oil, mineral, acrylic — that is matter. The work occupies a particular location: on a wall, in a room, at a distance from the viewer. That is space. The experience of seeing unfolds across seconds and minutes. First the whole image. Then the color. Then the figure. Then the detail you missed. Then the feeling that arrives late. That is time.

A painting is a four-dimensional encounter.

It looks still, but it is not still.

The viewer activates it.

Without light, the painting vanishes.

Without matter, it has no body.

Without space, it has no relation.

Without time, it cannot be received.

That is the architecture of everything compressed into a frame.

A sculpture makes this even more physical. Walk around it and the object changes without moving. The form reveals itself through your movement in space. One angle hides what another angle exposes. Light catches the surface differently as your position changes. The sculpture teaches you that truth is not always visible from one side.

That is not only art.

That is life.

A building is art you can enter. Architecture organizes matter into space for bodies moving through time under light. A good room changes your nervous system. Ceiling height alters breath. Window placement changes mood. Materials hold temperature. Sound behaves differently against wood, stone, glass, fabric. A chapel, a courthouse, a childhood kitchen, a hospital hallway — each one arranges the body differently because space is not neutral.

Space speaks.

Matter speaks.

Light speaks.

Time speaks through wear.

This is why old buildings feel different. Time has entered the matter. Edges softened by hands. Floors marked by footsteps. Wood darkened by sun. Stone polished by use. A new room may be clean, but an old room has memory embedded in surface.

Music is one of the clearest signs that reality is more than static matter.

It is vibration made meaningful: born in matter, carried through space, unfolding through time.

Wood, string, air, breath, hand — each submits to pattern, and pattern becomes feeling.

A guitar string vibrates. It moves air. The air presses against the eardrum. Tiny bones transmit the motion. Fluid moves in the inner ear. Hair cells bend. Electrical signals travel into the brain. The brain turns vibration into pitch, rhythm, memory, ache.

Music is physics entering grief and coming out as beauty.

But music belongs most deeply to the fourth dimension. Time is the true medium of sound. A painting occupies space. A sculpture occupies form. Music lives by passing away. It cannot be held still without ceasing to be itself.

Pause a song forever on one note and you have not preserved it.

You have killed it.

Music requires disappearance.

Each note must leave to make room for the next. The melody depends on loss. The chord changes. The rhythm moves. The chorus returns, but it returns changed because you have now traveled through the verse.

This is why music can break you open so quickly.

It teaches the body the structure of time.

Order.

Tension.

Memory.

Longing.

Release.

Music does not merely move through time.

Music makes time visible.

Film is all four dimensions simultaneously: light projected through matter across space over time.

The filmmaker creates a world. The actors populate it. The cinematographer frames perception. The editor organizes time. The viewer sits in darkness and receives a simulation of consciousness — a mind moving through space, encountering matter, carried by sequence, illuminated by real light.

A film is not merely watched.

It is entered.

The room goes dark, and the ordinary world withdraws. Light appears on a surface. Faces larger than life move before you. Sound enters the body through the chest and ears. Time is compressed, expanded, rearranged. A childhood can pass in ninety seconds. One glance can hold for ten. A death can be replayed. A memory can be cut into the present.

Film teaches us that consciousness is edited.

Your own mind does this constantly. It cuts from sensation to memory, from image to fear, from sound to association. It frames. It omits. It emphasizes. It scores the scene with chemistry. The world arrives whole, but you experience a version assembled through attention.

That is why film became the dominant art form of the modern age.

It resembles the machinery of perception.

But consider the parallel more carefully. A film requires

source code, a director, a frame, a cast, a setting, a timeframe, and a budget.

These are not just production terms.

They are metaphysical terms wearing baseball caps.

Source code: the laws of physics.

Director: the animating intelligence of the whole.

Frame: the perceptual limits of consciousness.

Cast: the beings inhabiting the simulation.

Setting: the environment where interaction occurs.

Timeframe: the sequence through which meaning unfolds.

Budget: the finite currency of attention, energy, and time available for the experience.

Every life has these.

Every day has these.

You wake into a production already underway. Light through the window. Body in bed. Memory online. Roles waiting. Constraints active. Time limited. The scene begins whether you feel prepared or not.

Then you improvise.

That is the terrifying and beautiful part.

The script is not fully written.

Art is not imitation of life.

Art is life recognizing its own operating system.

This is why standing before a painting that moves you produces something in the chest that is not quite pleasure and not quite pain. That specific ache of beauty. The Japanese call it mono no aware — the pathos of things, the tender sadness of impermanence.

The ache comes because the body recognizes the pattern.

The flower is beautiful because it will wilt.

The child is beautiful because they will grow.

The song is beautiful because it ends.

The light is beautiful because it is moving.

The moment is beautiful because it cannot be held.

Beauty is time touching matter through light in space and reminding you that nothing here stays fixed.

That is not depressing.

That is what makes it alive.

When art moves you, the universe is experiencing itself from two angles at once: through the mind that made the work and through the mind that receives it. The painter's hand is gone, but the stroke remains. The composer's breath is gone, but the note returns. The actor's moment was filmed years ago, but your body reacts now.

Time folds.

Space collapses.

Matter carries light.

Meaning crosses.

You are the universe looking at itself through a mirror it built from pigment, sound, body, and flame.

The ache is recognition.

I know this.

I am this.

How did I forget?

LIVING INTEGRATION: DIMENSIONAL AWARENESS

Take an ordinary object from your immediate environment — a cup, a pen, a stone, a leaf, whatever is nearest.

Hold it in your hand.

Do not rush.

Let the object have its full weight.

First, notice the light on it. Watch how illumination travels across its surface. Where does it brighten? Where does it fall into shadow? What color appears because certain wavelengths reflect into your eye? What texture becomes visible only because the light arrives from a specific angle?

This is the first dimension at work: light reaching the object, bouncing from it, traveling to your eye, allowing the world to appear.

Spend thirty seconds with only the light.

You are watching the universe's first language speak.

Next, notice the space it occupies. Its height relative to your fingers. Its width. Its distance from your chest. The air around it. The empty space that outlines it. The gap between the object and the wall, between the object and the floor, between the object and your face.

Space is not nothing.

Space is relation made available.

The object is not isolated. It exists in a web of distances. Move it closer and intimacy changes. Move it farther and perception changes. Put it behind glass and access changes. Place it in someone else's hand and meaning changes.

Spend thirty seconds attending to where the object is.

Now feel its matter.

The weight. The density. The temperature. The resistance. The way it pushes back against your grip. Too light and it would feel insubstantial. Too heavy and the hand would strain. Its matter is not an idea. It is a fact pressing into your palm.

This is the third dimension: form with consequence.

Matter holding shape.

Light staying long enough to be touched.

Spend thirty seconds feeling the materiality of the object.

Finally, notice time.

This moment — you holding this object, in this room, in this body — is occurring now and only now. A moment ago, you were reading. In a moment, you will lower your hand. The object may remain, but this exact contact will pass. The warmth from your hand may linger briefly on its surface. Then even that will fade.

Time is moving through the object and through you.

It is allowing the encounter to happen, and it is also taking it away.

Spend thirty seconds attending to the temporal dimension.

You just held the architecture of reality in your hand.

You do this constantly.

With every cup, every key, every door handle, every hand you hold, every face you see, every meal, every room, every ordinary object you stop noticing because the mind has labeled it too quickly.

The universe is always fully present.

Light.

Space.

Matter.

Time.

The only thing that changes is your attention.

CHAPTER FOUR

THE BODY ELECTRIC
Instruments of the Infinite

"Our bodies are more than containers for the soul. They are living instruments through which the universe becomes felt, heard, and known."

NODE IN THE NETWORK

Place your hand on your chest.

Not symbolically.

Physically.

Feel the heel of the palm settle against the ribs. Feel the skin warming under contact. Feel the small rise and fall of breath beneath the hand. If you wait long enough, you feel the heartbeat—not always clearly at first, sometimes faint, sometimes irregular in its visibility, but there. A pulse. A pressure wave. A muscular contraction sending blood through the branching rivers inside you.

That pulse is not an idea.

It is a signal.

Your body is speaking in rhythm before the mind has formed a word.

Your body is a node.

I spent years in network engineering before the universe redirected my path toward the larger project that produced these pages. In that earlier life, I thought about nodes constantly. I thought about how information flows through networks. About the difference between a node and a conduit. About what makes a network intelligent rather than merely connected.

A conduit carries.

A node interprets.

That distinction matters.

A simple cable allows signal to pass through. It does not know what the signal means. It does not decide whether to store it, amplify it, reject it, route it differently, or integrate it with other streams of information. It conducts, but it does not understand.

A node is different.

A node receives.

A node processes.

A node transmits.

A node changes the system by participating in it.

In a technical network, a node can be a router, a switch, a server, a terminal, a device with an address, a place where information arrives and is handled according to rules. It receives packets. It reads headers. It checks integrity. It decides where the information goes next. A network becomes intelligent when its nodes do more than pass signal forward. They interpret. They regulate flow. They detect failure. They adapt.

Your body is a node of extraordinary sophistication.

Not a metaphorical node.

A living one.

Every second, your nervous system receives an enormous stream of sensory information from the environment and from inside the body. Light enters the eyes. Sound presses against the eardrums. Skin detects texture, temperature, pressure, vibration. Proprioceptors tell you where your limbs are without requiring you to look. Interoceptors monitor hunger, heartbeat, breath, gut tension, blood pressure, internal state. The vestibular system tells you how your head is positioned in gravity. Your body is continuously mapping the boundary between self and world.

Most of it never becomes conscious.

That should humble you.

The part of you that says "I" is not running the whole operation. It is receiving a highly edited dashboard.

Behind the dashboard, the system is enormous.

Your conscious mind handles a narrow band of information. Beneath that narrow band, the body is processing millions of details: balancing posture, regulating temperature, scanning faces, reading tone, adjusting hormones, interpreting smell, detecting threat, coordinating digestion, repairing tissue, calibrating immune response, storing memory, dreaming, anticipating, orienting.

You are not a single thought riding around in a body.

You are a civilization of signals.

Your heartbeat is information. Your posture is information. Your breath rate is information. Your facial muscles transmit information before language arrives. Your gut tightens before your mind can explain why a room feels wrong. Your shoulders rise before you admit you are afraid. Your body knows things before the narrator catches up.

That is not irrational.

That is bandwidth.

The nervous system is not waiting for your permission to process reality. It is already doing it. It has always been doing it. It began long before you had language and continues every night when language dissolves into sleep.

You are not merely in the universe.

You are a processing node of the universe.

The universe is not simply happening to you. It is flowing through you as light, sound, pressure, heat, gravity, breath, food, memory, contact, and emotion. Your nervous system receives that flow, organizes it, makes meaning of it, and transmits something back through speech, action, expression, field, choice, and presence.

A router does not create the internet.

But the internet cannot function without routers.

A musician does not create the laws of sound.

But music cannot happen without an instrument.

A window does not create sunlight.

But the room is changed by the way the window receives it.

You are like that.

A receiver.

A processor.

A transmitter.

An instrument.

When you feel the warmth of sunlight, you are not merely receiving a resource. You are completing a circuit. The sun generated light. That light crossed space. Your skin received it. Your biology responded. Your mood shifted. Your posture changed. Your own body emitted heat, reflected light, altered breath, altered chemistry. Something entered you, and something left you changed.

That is a circuit.

The circuit runs through you.

This is what most modern people miss. They imagine the body as a private object moving through a separate world. A sealed bag of organs having personal experiences. But no body is sealed. You are porous at every level. You inhale the atmosphere. You absorb light. You exchange heat. You take in food that used to be soil, sun, water, and life. You release carbon dioxide that plants will use. You carry microbes that help digest your meals. You are touched by other people's voices, scents, moods, and silences.

The boundary is real.

It is not absolute.

Skin is not a wall.

Skin is an interface.

The body is more like a coastline than a container. It has shape. It has boundary. But it is constantly being met, changed, shaped, and rewritten by the waves.

We play the song of our lives with this instrument. The universe composed the laws. We perform within them. And this is the extraordinary part: our performance is not the reproduction of a fixed score. It is improvisation. The universe gave us the instrument, the chord structure, the tempo of mortality, the physics of sound, the limits of breath, and left the melody to us.

That is freedom.

Not freedom from structure.

Freedom through structure.

A guitar string can only make music because it is constrained.

It is fixed at both ends. It has tension. It has length. It has material properties. Without constraint, it produces no note. The same is true of a human life. The body has limits. Time has limits. Gravity has limits. But within those limits, tone becomes possible.

You are not trapped in the instrument.

You are being given one.

LIQUID CRYSTALS

Here is an idea that changed how I understand everything about the human body, and it is more biology than philosophy:

The human body behaves in many ways like a liquid crystal.

Not literally as one uniform crystal, but as a structured, fluid, electrically active biological system.

This requires unpacking.

A crystal is a solid whose atoms or molecules are arranged in an ordered, repeating pattern. That order gives crystals properties that disordered materials do not have. The shape matters. The internal arrangement matters. Pressure, charge, orientation, and vibration can move through ordered structures in precise ways.

Quartz teaches this clearly.

Squeeze a quartz crystal and it generates electrical voltage. Apply voltage and it changes shape. This is piezoelectricity — the conversion of mechanical pressure into electrical charge and electrical charge into mechanical movement. That principle lives inside quartz watches, microphones, ultrasound devices, sensors, and countless instruments that translate one kind of signal into another.

Pressure becomes voltage.

Voltage becomes motion.

Motion becomes information.

That is not mystical.

That is engineering.

Now bring that principle back to the body.

The body is not a pile of wet parts. It is structured. Ordered. Conductive. Responsive. It is made of tissues that carry charge, fluids that move ions, membranes that maintain voltage differences, bones that respond to stress, fascia that transmits tension, nerves that fire electrical impulses, heart tissue that contracts through electrical coordination.

You are not just chemical.

You are electrical.

You are mechanical.

You are fluid.

You are structural.

And all of those systems are speaking to each other at once.

The human body is mostly water, but not all water in the body behaves like water in a glass. Biological water exists in crowded environments: near membranes, proteins, collagen fibers, DNA, minerals, and cellular surfaces. In those environments, water can become more ordered than ordinary bulk water. It participates in charge separation, molecular organization, transport, and the subtle architecture of living tissue.

The body is not a bottle full of loose liquid.

It is a living matrix saturated with organized fluid.

Fascia makes this easier to feel. Press your thumb into the palm of your other hand and slowly drag it across the tissue. You can feel resistance. Stretch. Sliding layers. Not just skin, not just muscle, but a connective web running through the body, wrapping organs, linking muscle groups, shaping posture. Fascia is not packing material. It is a communication system made of tension, hydration, collagen, and motion.

Pull on one corner of a shirt, and the fabric shifts somewhere else.

Pull on fascia, and the body answers elsewhere.

That is tensegrity.

The body is built like a suspension bridge, a sailboat rig, a spiderweb, a geodesic dome. Compression and tension working together. Bone as strut. Fascia as cable. Muscle as adjustable tension. The whole structure distributes force across the whole system.

This is why one tight jaw can change the neck.

Why one injured foot can alter the hip.

Why grief can live in the shoulders.

Why posture is not merely mechanical, but emotional history held as shape.

Your body remembers by organizing itself.

Not only in the brain.

Everywhere.

Bones also respond to mechanical stress. Load them and they remodel. The skeleton is not dead scaffolding. It is alive, vascular, responsive. Walk, lift, carry, press, and the bone receives instruction through force. It strengthens along lines of demand. Remove load, and bone thins. The structure listens.

Pressure writes into matter.

This is why the body cannot be understood as a passive vessel. It is an adaptive instrument. It receives force and changes shape. It receives light and changes chemistry. It receives sound and changes rhythm. It receives touch and changes state.

A violin teaches the same lesson. The wood matters. The curve matters. The tension of the strings matters. The hollow chamber matters. The player matters. Press the string in the wrong place and the note sharpens or dies. Draw the bow with too much force and the sound breaks. The instrument is sensitive because it is ordered.

Sensitivity is not weakness.

Sensitivity is precision.

The body is precise.

Thoughts are part of that precision. Not as magical commands floating in space, but as patterns of neural firing, hormone release, muscular tension, breath change, attentional orientation, and behavioral preparation. A thought is not "just a thought." It is an event in a biological electrical system.

Think of someone who betrayed you.

The body changes.

Jaw tightens. Gut contracts. Breath shortens. Heat rises. Eyes narrow. Shoulders armor.

Now think of someone who makes you feel safe.

The body changes again.

Breath drops. Face softens. Chest opens. The abdomen releases. Hands unclench.

Nothing outside the body changed.

But the instrument changed state.

That is the point.

Your emotions are not vapor. They are not private clouds sealed inside the skull. They are embodied patterns. Fear is not only an idea; it is adrenaline, cortisol, pupil dilation, muscle readiness, blood flow redistribution, vigilance, narrowed attention. Gratitude is not only a polite thought; it is breath, warmth, softened facial tone, altered heart rhythm, wider perception, social openness.

Emotion is body weather.

It changes pressure, temperature, charge, and motion.

This is why the phrase "good vibes" is sloppy but not entirely wrong. People feel state. A tense person transmits tension through face, voice, timing, breath, posture, word choice. A grounded person transmits stability through those same channels. The body reads the room faster than the mind can write the report.

Chronic resentment is not just a story you tell yourself. It becomes repeated physiology. Repeated physiology becomes pattern. Pattern becomes posture. Posture becomes pain. Pain becomes identity if left unquestioned.

The body keeps receipts.

Joy works the same way in the other direction. Sustained joy, sincere gratitude, meaningful connection, laughter that moves through the belly — these are not decorative experiences. They alter chemistry, breath, tone, immune function, behavior, and the quality of attention. The body becomes more coherent because the internal signals are no longer fighting each other.

A tuned instrument rings.

An untuned instrument still makes sound, but the sound argues with itself.

How do you charge a crystal?

You expose it to light. You place it in contact with minerals. You apply pressure. You vibrate it. You align it. You treat its structure as something capable of receiving and transmitting.

How do we charge the body?

Water. Minerals. Sunlight. Breath. Motion. Rest. Touch. Focus. Meaning.

Potassium, sodium, magnesium, calcium — these are not

wellness decorations. They are ions. They carry charge. Nerve impulses and muscle contractions depend on gradients of charged particles moving across membranes. The heart depends on electrical timing. The brain depends on electrochemical signaling. Hydration and minerals are not luxuries for the instrument. They are part of the wiring.

Sunlight is not only mood.

It is timing.

Morning light calibrates the circadian system. Darkness allows sleep chemistry. Sleep repairs the instrument. Movement circulates fluid, loads bone, pumps lymph, clears metabolic waste, organizes fascia. Breath shifts carbon dioxide and oxygen levels, changes heart rhythm, modulates nervous system state.

This is the practical spirituality of the body.

Drink water.

Get sunlight.

Move.

Breathe.

Touch the Earth.

Tell the truth.

These are not small things.

They are maintenance instructions for a living instrument.

THE HANDS THAT KNOW

Look at your hands.

Turn them over.

Palms up.

Then palms down.

The hands are so familiar they become invisible. But they are among the most extraordinary instruments you possess. Each hand is a field of sensors, bones, tendons, muscles, blood vessels, nerves, heat, memory, and intention. They can crack open a coconut, thread a needle, hold a newborn, play a chord, write a sentence, stop a fall, bless a forehead, grip a steering wheel, bury the dead, build a home.

Hands are where thought becomes world.

Before a house exists, hands lift material.

Before music enters a room, hands touch strings, keys, drum, fader.

Before love becomes visible, hands reach.

The hands contain dense tactile innervation. They detect texture, pressure, temperature, vibration, movement. They can distinguish subtle differences in surface. They know the grain of wood, the slickness of oil, the brittleness of dry leaves, the fever heat of a child's forehead.

The hands think.

Not with words.

With contact.

Touch a surface with your fingertips and the body immediately begins mapping. Smooth or rough. Safe or sharp. Warm or cold. Familiar or strange. The information travels fast. The hand reports to the nervous system, and the nervous system updates reality.

This is why children touch everything.

They are not being irrational.

They are building the world.

But hands are not only receivers.

They are transmitters.

When you touch another person, you are not merely placing skin against skin. You are establishing a biological exchange. Heat transfers. Pressure communicates. Nerves fire. Hormones shift. Breath responds. Muscles soften or guard. The body decides whether the touch is safe, intrusive, loving, mechanical, threatening, healing.

Touch tells the truth quickly.

A forced hug and a safe hug are not the same event.

The body knows.

A hand placed on the shoulder at the right moment can do what language cannot. It can say: I am here. You are not alone. Stay in your body. Keep breathing. The pressure gives the nervous system an external point of organization. It helps the body find itself.

This is why touch is healing.

Not merely emotionally.

Physiologically.

Safe touch can reduce stress, support bonding, calm the nervous system, and change the felt state of the body. A child held against a parent's chest receives warmth, rhythm, scent, pressure, and regulation. An elder whose hand is held receives proof of contact. A grieving person may not need advice. They may need a hand.

Words can fail at the edge of pain.

Touch still speaks.

The healer's hands, the mother's hands, the musician's hands, the builder's hands, the surgeon's hands, the gardener's hands — all of them translate invisible intention into physical contact. They bring the inner state into matter.

This is why traditions use hands in blessing. The laying on of hands. Hands lifted in prayer. Hands folded. Hands open. Hands placed over the heart. Hands washed before sacred work. Hands carrying bread, incense, water, oil, child, tool, instrument.

The gesture matters because the body understands gesture.

The hand is not symbolic after the fact.

The hand is a signal.

And this is why the ring carries significance that transcends culture.

Across civilizations, the ring appears in ceremonies of commitment. A circle of metal placed on the finger. Conductive material wrapped around one of the most sensitive and expressive parts of the body. Gold. Silver. Platinum. Metal shaped into continuity.

The ring is not powerful because culture randomly assigned it power.

Culture assigned it power because the form already speaks.

Circle: no beginning, no end.

Metal: durability, conductivity, value, resistance to decay.

Finger: touch, action, promise entering daily use.

A ring sits where the hand meets the world. It moves when you work, cook, write, lift, hold, build, comfort. It is commitment made tactile. A promise that touches the body all day long.

The ring reminds the nervous system.

Chosen connection.

Maintained relation.

A circle worn where intention becomes action.

We know this without being taught.

Something in us understands.

The universe put that understanding in the body before culture gave it ceremony.

CHARGING THE HUMAN CRYSTAL

If the body behaves like a liquid crystal instrument, then tuning matters.

Charging matters.

Input matters.

You cannot feed the instrument static all day and expect it to sing clearly.

This is blunt because it needs to be blunt.

Your thoughts are not private in the way you imagine. They may not be audible to another person, but they are not sealed away from the body. Every repeated thought becomes repeated chemistry, repeated posture, repeated breath, repeated attention, repeated behavior.

A thought held once is a spark.

A thought held for years is a climate.

You live inside the weather you keep generating.

The primary charging mechanism is sustained, emotionally vivid, focused attention that changes the state of the body and organizes action.

The body knows the difference between real gratitude and fake gratitude.

It knows the difference between peace and collapse.

It knows the difference between forgiveness and avoidance.

It knows the difference between desire and desperation.

You cannot lie to the instrument for long.

The face will show it. The breath will show it. The gut will show it. The sleep will show it. The tone of voice will show it. The body is the final auditor.

Psychoneuroimmunology has already shown that

psychological states influence nervous system activity, hormonal patterns, inflammation, immune response, and recovery. Stress is not merely a feeling. It becomes measurable strain. Chronic fear loads the system. Chronic isolation leaves marks. Meaning, connection, and hope also leave marks.

The mind and body are not two kingdoms exchanging letters by courier.

They are one system described from different angles.

Heart rate variability gives one window into this. A healthy heart does not beat like a metronome. It varies. It adapts. It responds to breath, movement, emotion, demand. Flexibility is a sign of intelligence. Rigidity is danger.

This is true everywhere.

A rigid mind breaks.

A rigid body tears.

A rigid life cannot absorb change.

Coherence is not stiffness.

Coherence is organized flexibility.

The most powerful states are not loud. They are sincere. Genuine gratitude. Real love. Authentic appreciation. A felt sense of safety. A clear intention connected to action. These states organize the instrument because they reduce internal contradiction.

Performed happiness does not do this.

Forced positivity does not do this.

Spiritual bypassing does not do this.

Painting a smile over a wound is not healing.

The body knows what is under the paint.

This is where discipline enters. Not harsh discipline. Not punishment. Tuning.

A musician tunes before playing because small deviations matter. A string slightly sharp changes the chord. A drumhead too loose changes the groove. A microphone placed wrong changes the recording. The audience may not know the technical reason, but they feel the difference.

Your life is the performance.

Tune the instrument.

The variables that determine the power of a wave also help us understand the power of thought: duration, intensity, coherence, repetition.

How long do you hold the pattern?

How vividly does the body participate?

How often do you return to it?

How much contradiction is present in the system while you hold it?

A single positive thought passing through a chaotic body is real, but brief. A ripple. It moves and disappears.

A sustained intention held with emotional clarity, repeated over time, supported by behavior, becomes a standing wave. It begins to organize perception. You notice opportunities that match it. You make decisions that support it. You speak differently. You stand differently. You move toward different people and away from different rooms.

This is where people misunderstand manifestation.

They want the universe to behave like a vending machine.

Insert thought.

Receive object.

That is childish.

The deeper mechanism is more demanding and more beautiful. Attention changes state. State changes behavior. Behavior changes environment. Environment changes probability. Probability changes outcome. The field responds not to fantasy alone, but to coherent participation.

The magnet does not choose its field.

It is its field.

You are the same unless you become deliberate.

Be clear and deliberate about your thoughts. They are not harmless entertainment scrolling through the sealed room of your skull. They are instructions entering tissue. They are rehearsals for action. They are patterns in the instrument.

Think resentment long enough and you practice resentment.

Think courage long enough and you practice courage.

Think scarcity long enough and you scan for lack.

Think gratitude long enough and you begin to detect support.

This is not moral judgment.

It is training.

The nervous system becomes efficient at what it repeats.

That is the gift and the danger.

LIVING INTEGRATION: TUNING THE INSTRUMENT

Place both hands over your heart.

Feel the pressure of your palms against the chest. Feel the warmth building between hand and body. Let the elbows drop. Let the shoulders soften. Let the back of the neck lengthen slightly. Let your jaw unclench.

Do not perform peace.

Just stop interrupting the body for a moment.

Feel the rhythm beneath your hands. This fist of muscle has been contracting and releasing since before you were born. Before you had a name. Before you had language. Before you had memory. The heart was already moving.

It has carried you through every room you have ever entered.

Every fear.

Every birthday.

Every humiliation.

Every kiss.

Every loss.

Every ordinary morning you forgot to notice.

It kept going.

Not because you managed it.

Because life was faithful inside you.

Now bring to mind something you love without reservation.

A person. A place that makes your body exhale. The ocean at a certain hour. A dog sleeping near your feet. A memory where you were not performing, not proving, not defending. A moment that lives in the body as warmth.

Do not force the feeling.

Let it come by proximity.

Like standing near fire.

Let the image or memory sit in the chest and let the body respond.

Notice what happens.

Warmth spreading.

Breath deepening.

Face softening.

A small ache behind the eyes.

A release in the shoulders you did not know were lifted.

The body recognizes love before the mind explains it.

Stay there.

This is not only psychological. It is breath, muscle, heart rhythm, electrical signaling, chemistry, memory, meaning, and attention entering a more coherent pattern. You are adjusting the instrument with focus.

Not fantasy.

Focus.

Now notice the difference between thinking about love and feeling love. Thinking stays mostly in the head. Feeling changes the chest, throat, breath, eyes, hands, belly. The body participates.

That is the doorway.

Practice this deliberately for five minutes each morning before anything else. Before the phone. Before the messages. Before the world begins inserting its agenda into your nervous system.

Hands on heart.

Breath felt.

Love remembered.

Instrument tuned.

Not as a technique for getting what you want.

As a practice of remembering what you are.

You are an instrument of the infinite.

Instruments require tuning.

The tuning is not work.

The tuning is coming home.

CHAPTER FIVE

EGO, ENERGY, ARCHIVE

"Ego is not the enemy. It is the hard drive. The question is: what have you stored on it?"

THE NECESSARY SELF

Feel your name on the tip of your tongue.

Say it silently.

Notice what happens.

The mind does not treat your name like any other word. It turns toward it. It gathers history around it. Faces appear. Rooms appear. Old versions of you come forward. The child. The teenager. The person who wanted to prove something. The person who was hurt. The person who survived. The person who built defenses so carefully they began to feel like personality.

Your name is not only a label.

It is a folder.

Open it, and the archive wakes up.

Let me say something about ego that most spiritual books are reluctant to say, because many spiritual books are written from the perspective that ego is the obstacle — the thing standing between the practitioner and enlightenment, between the seeker and their destination.

Ego is not the obstacle.

Ego is the vehicle.

The problem is not that you have an ego. The problem is that your ego may have forgotten that it is a vehicle and convinced itself that it is the destination.

That is the trouble.

Not the self.

The self mistaking itself for the whole.

A car is not a prison unless you forget you are driving it. A helmet is not a problem unless you refuse to take it off when the danger has passed. A password is useful until it locks you out of your own house.

The ego is like that.

Necessary.

Protective.

Limited.

Dangerous only when promoted beyond its role.

Ego is an energy archive. Think about this carefully, because the metaphor is not decorative. It is structural. Your ego is the sum total of every experience you have encoded into identity. Every time something happened to you and you drew a conclusion about yourself — I am the kind of person who does this, I am not the kind of person who does that, this is safe, this is dangerous, this is who I am under pressure, this is what I need to feel okay — that conclusion entered the archive.

Layer by layer.

Experience by experience.

Impact by impact.

The archive compiled itself into the complex structure that introduces itself when someone asks your name.

A child touches a hot stove and the body stores heat as danger. That is intelligence. A child reaches for affection and receives rejection, and the system stores reaching as risk. That is also intelligence, but now the archive begins to build identity around pain. A teenager gets laughed at in a moment of vulnerability and the ego writes a rule: never expose that part again. An adult succeeds after years of pushing and the ego writes another rule: rest is unsafe; value must be earned through strain.

The archive is always writing.

The question is whether anyone is reviewing the file.

This archive is not your enemy. It is your inheritance. It contains your navigation system for the world you have actually lived in. It contains emotional memories that tell you, faster than thought, that this situation resembles a situation that hurt you before, or that this person carries a quality you have learned to trust. It contains aesthetic preferences that make you who you are: the music that opens your chest, the specific quality of light that feels like home, the smell that takes you back twenty years before the mind can stop it.

Without the archive, you would face every moment as if it were the first moment of your existence.

You would not know your own hand.

You would not know which voice belongs to your mother.

You would not know that fire burns, that betrayal has a taste, that certain rooms make your shoulders tighten, that certain songs bring the dead close.

The archive gives continuity.

Continuity gives self.

Self gives responsibility.

We need the ego.

The problem begins when the ego needs to be the most important thing in every room it enters.

You have felt this in the body. Ego does not remain theoretical. It changes posture. When the ego is threatened, the chest tightens. The jaw sets. The throat constricts. The stomach drops or hardens. The breath rises higher. The eyes scan. The body prepares to defend not flesh, but image.

That is ego's great confusion.

It treats image as survival.

It treats correction as attack.

It treats being misunderstood as death.

It treats losing status like losing oxygen.

This is not because ego is stupid. It is because ego evolved around protection. The system learned that belonging matters, reputation matters, approval matters, hierarchy matters, rejection hurts, exile can kill. The ancient body remembers that separation from the group once meant real danger. So the ego watches the social field like a guard at the gate.

The guard is useful.

Until it starts running the kingdom.

THE ARCHITECTURE OF SELF

The ego expresses itself through four primary channels, and understanding these channels is the beginning of working with the ego intelligently rather than being unconsciously driven by it.

The first channel is self-reference.

At the most fundamental and necessary level, ego is the

capacity to distinguish self from other: my hunger is my hunger and your pain is your pain; I inhabit this body and you inhabit that one; I have preferences and you have different preferences. This is not pathology. This is the basic equipment of a self-aware organism moving through a world of other self-aware organisms.

Without self-reference, you cannot act with intention.

You cannot choose.

You cannot take responsibility.

Self-reference is foundation.

The body teaches this before language. An infant learns boundary through touch, hunger, warmth, absence, return. The breast is not the mouth. The hand is mine. The face is yours. The cry brings someone. The cry does not bring someone. Slowly, through contact and absence, the self begins to form.

The ego begins as orientation.

Where am I?

What is mine?

What is not mine?

Where do I end?

Where does the world begin?

A GPS needs a current location before it can route anywhere. A house needs a front door. A body needs skin. A self needs enough boundary to move through the world without dissolving into every surrounding signal.

So self-reference is not the enemy.

But it can thicken.

It can become obsession.

Instead of "I am here," it becomes, "How am I being seen here? Am I important here? Am I safe here? Am I winning here? Does this room understand who I am?"

That is when orientation becomes performance.

The second channel is pride.

Having established a self, the ego immediately wants to know: how does this self compare to other selves? Is it adequate? Is it sufficient? Is it more than, less than, equal to? Pride is the ego's measurement function — the constant, automatic process of assessing one's relative position in the social and existential field.

Pride at its best is dignity.

The spine standing up.

The refusal to be treated as worthless.

The inner line that says: I am not above you, but I am not beneath you either.

That kind of pride is clean. It protects the human being from collapse. It allows a person to take up space without apology. It gives the voice enough ground to speak.

Pride at its worst is hunger wearing armor.

The compulsive need to be better than. To win. To be seen winning. To accumulate evidence of superiority because the actual experience of worth is missing or unstable.

That kind of pride never rests.

It cannot.

The measurement never ends.

There is always someone richer, younger, more beautiful, more accomplished, more admired, more spiritual, more disciplined, more free. Pride that depends on comparison has to keep scanning because its food is always outside itself.

This is exhausting.

You can feel it in the body.

The lean forward. The tight smile. The need to insert achievement into conversation. The inability to celebrate someone else without quietly checking what their success means about you. The little internal flinch when another person shines.

That is not evil.

That is pain trying to become status.

The third channel is comparison.

The algorithm of modern suffering.

When social media platforms engineered themselves into the most effective delivery mechanisms for comparison that human culture has ever produced, they did not accidentally stumble into addiction. They designed systems that manipulate the measurement function of the ego: infinite scroll, variable reward, public metrics, curated images, filtered bodies, visible approval, invisible context.

A slot machine for self-worth.

Pull down.

Refresh.

Check.

Compare.

Repeat.

Your nervous system was not built to compare your private interior to the polished exterior of thousands of people per day. It was not built to watch strangers perform success, beauty, outrage, family, wealth, grief, virtue, and enlightenment in rapid succession while your body sits under artificial light holding a rectangle.

That is not connection.

That is psychic abrasion.

Social networks engineer feeling through words, images, audio, timing, friction, reward, outrage, desire, fear, envy, belonging. They do not merely show you the world. They shape the state from which you encounter the world. They teach the ego what to want, what to fear, what to resent, what to perform.

When you surrender the controls of your emotional frequency to an external system that profits from dysregulation, you give away one of the most important things you possess: the ability to feel what you actually feel in response to what is actually happening in your actual life.

You trade direct experience for mediated comparison.

You stop tasting your own meal because you are busy looking at photos of someone else's dinner.

You stop feeling your own home because someone else's house looks brighter.

You stop hearing your child laugh because a stranger's vacation seems more cinematic.

This is theft.

Not of money.

Of presence.

The fourth channel is desire.

The ego wants. Status, security, sensation, significance. More of the good, less of the bad, more control over which is which. It wants praise. It wants proof. It wants certainty. It wants to be chosen. It wants to avoid humiliation. It wants the body safe, the future secured, the identity validated, the wound compensated.

Desire itself is not the problem.

Desire is ancient.

Desire is the engine of creation. The reaching from zero toward one. Without desire, nothing moves. No seed splits. No child reaches. No artist begins. No scientist asks. No lover crosses the room. No prayer rises.

Desire is life leaning forward.

The problem begins when desire becomes attachment.

When wanting becomes needing.

When reaching becomes grasping.

When the ache of incompleteness that drives growth hardens into the conviction that you cannot be okay until the desired thing is yours.

The body knows this difference.

Clean desire feels alive. It has heat, direction, movement. It opens the chest and sharpens attention.

Attachment feels tight. The jaw clenches. The breath shortens. The hand closes. The mind loops. The desired object stops being something you want and becomes the imagined condition of your survival.

That is the trap.

Desire reaches.

Attachment clutches.

One participates in creation.

The other tries to control it.

A surfer understands this. You paddle for the wave with intention. You commit. You read the water. You move. But if you try to grip the wave, dominate it, force it to become the ride you imagined, you fall. The wave has its own structure. You participate or you get worked.

Desire is paddling.

Attachment is trying to own the ocean.

MIND ENERGY: THE UPGRADE

There is a higher operating system available.

Every major wisdom tradition has pointed at it from a different angle, using different names, describing slightly different features of the same fundamental upgrade. I call it mind energy — not because the name is perfect, but because it captures the essential idea: the power of awareness itself, prior to the archive, prior to ego commentary, prior to the self that the ego has constructed.

Mind energy is what remains when you subtract the commentary.

Sit quietly and watch a thought appear.

Not the content. The appearance.

A sentence forms in the mind. A worry. A memory. A plan. An old argument you are still winning in the shower years later. It appears, occupies attention, changes, fades, returns, mutates.

Now ask: who sees it?

The thought is not seeing itself.

There is awareness in which the thought appears.

That awareness is not loud. It is not dramatic. It does not need a costume. It is the clear space in which the archive displays its files. It is the screen on which the film plays, not any particular scene in the film.

That witnessing awareness is what I mean by mind energy.

The ego says, "This thought is me."

Mind energy says, "This thought is appearing."

That small difference is liberation with work boots on.

Not floating away.

Not becoming special.

Just no longer being dragged by every mental weather pattern that crosses the sky.

Mind energy's qualities are the polar complements of ego's four channels.

Where ego offers self-reference — the continuous measurement of I against everything — mind energy offers truth.

Not my truth as preference.

Not your truth as opinion.

Truth as what is actually happening before the story arrives.

The sky is grey and rain is falling.

That is truth.

The sky is grey and my day is ruined.

That is ego commentary.

The body is tired.

That is truth.

I am weak because I am tired.

That is archive.

A person did not respond to my message.

That is truth.

They do not respect me, I am being rejected, I should protect myself, I always get ignored.

That is the ego opening old folders and throwing paper everywhere.

Mind energy can see the rain.

Ego experiences the story about the rain.

Where ego offers pride — the compulsive measurement of relative worth — mind energy offers understanding.

Understanding is the capacity to see without immediately ranking. To witness a person in their complexity without instantly sorting them into above or below, threat or resource, better or worse, useful or useless.

Understanding does not remove discernment.

It refines it.

You can still notice danger. You can still make decisions. You can still leave the room. But you are not driven by the automatic measurement function every second. You are no longer required to turn every person into a mirror for your own worth.

That is freedom.

A carpenter does not hate the wood for having knots. He reads the grain and works with what is there. A sailor does not take the wind personally. She trims the sail. A doctor does not insult the fever. She reads it as information.

Understanding reads.

Pride ranks.

Where ego offers comparison — the anxious surveying of other selves — mind energy offers acceptance.

Not passive resignation.

Not weakness.

Not pretending injustice is fine.

Acceptance means reality is allowed to be real before you decide how to respond.

That is power.

If your house is on fire, acceptance does not mean sitting in the living room pretending flames are sacred. It means you stop arguing with the fact of fire and move. Resistance wastes seconds insisting the fire should not be happening. Acceptance grabs the child and exits.

Acceptance is not surrendering action.

Acceptance is surrendering delusion.

This moment is what it is.

Now respond.

Where ego offers desire — reaching, grasping, needing — mind energy offers stillness.

The return to zero.

Not the death of desire, but its source.

The silence that holds every sound.

The capacity to want without needing, to reach without grasping, to love without possessing.

Stillness is not passivity. Stillness is the pause before clean movement. It is the archer's breath before release. The musician's silence before the first note. The builder's level before the wall goes up. The surfer watching the horizon before paddling.

Stillness is zero.

Creation is one.

Desire is the motion between them.

The art of living is learning when to reach and when to return. When to generate and when to be quiet. When to speak and when to listen. When to play the note and when to honor the rest that gives the note meaning.

SURRENDER PRECEDES STILLNESS

There is a moment in every life — often many moments, spread across years and decades — when the ego encounters something it cannot archive.

Something too large.

Too sudden.

Too final.

Too honest.

The death of someone essential.

The end of a marriage that was supposed to be permanent.

The diagnosis that restructures every plan.

The failure of the thing you organized your identity around succeeding at.

The betrayal you did not see coming.

The humiliation that strips the costume off in public.

The moment when the careful architecture of self that the ego has been building since childhood meets an event the architecture cannot hold.

You can feel it in the body.

The floor drops.

Breath disappears.

The chest becomes stone.

The mind searches for a file and finds none.

No category.

No plan.

No familiar script.

This is where ego suffers most: not in pain itself, but in pain it cannot control with identity.

In that moment, there is a choice.

Cling to the archive, protect the structure, rebuild as quickly as possible, restore the familiar operating system, and get back to normal.

Or let go.

Letting go is not elegant when it first arrives.

Often it comes after exhaustion. After bargaining fails. After anger burns through its fuel. After every strategy has been tried. After the ego has run all night through the burning house looking for a door that is no longer there.

Then something releases.

Not because you became wise.

Because you became tired enough to stop gripping the hot metal.

When the moment of letting go finally unfolds, it is not what fear imagined.

Fear said letting go would be death.

It was relief.

A wave through the body. Shoulders dropping. Breath returning. Hands opening. The nervous system realizing that the thing it had been clenching could be released without the entire self disappearing.

Once you let go and realize that it is okay, a giant wave of relief sets in. A rush of acceptance and divine understanding floods the body. It does not destroy the ego. It does not erase the archive, the identity, the self, the memories, the responsibilities. It dissolves rigidity. It softens the hard drive from a defensive, armored structure into something more fluid, more responsive, more alive.

The ego is not destroyed by surrender.

It is clarified.

Like a clenched fist carrying a stone for so long it forgot it was clenching. When the stone is released, the hand does not disappear.

It opens.

It becomes useful again.

The instrument recovers its range.

This matters when we speak of death.

I may still fear the pain of loss. Fear of physical pain is natural; it is the body's alarm system doing its work. The body is designed to protect form. It should flinch from the blade, pull back from fire, brace against impact, cry out when injured. That is not spiritual failure. That is biology preserving the vessel.

But I do not fear where I am going beyond that.

The return to oneness — the reality toward which wisdom traditions point from different angles — strikes me as beautiful. The difficulty is not the return. The difficulty is the clenching before it. The insistence that the wave must remain wave forever and never return to ocean.

But the wave was never separate.

That is the secret.

Perhaps heaven is our word for accepting that return.

Perhaps it is not only a place after death, but a state available whenever the fist opens. Whenever the ego stops mistaking tension for safety. Whenever the body releases its argument with what is.

Heaven as unclenching.

That is not small.

That is everything.

LIVING INTEGRATION: THE ARCHIVE INVENTORY

Take a piece of paper.

Draw a vertical line down the center.

On the left side, write:

EGO ARCHIVE

On the right side, write:

MIND ENERGY

Do not make this pretty.

Make it true.

Under Ego Archive, write five beliefs you hold about yourself. Not the polished ones. Not the ones you would put in a bio. The automatic ones. The ones that run beneath behavior. The ones formed by comparison, old hurt, early praise, humiliation, abandonment, pressure, survival, or what someone told you before you were old enough to question it.

Write the files that still open by themselves.

I am not enough.

I have to prove myself.

If I rest, I fall behind.

If people really see me, they will leave.

I must be useful to be loved.

I cannot trust good things to last.

I have to win to be safe.

Write yours.

The archive cannot be updated if you refuse to look at what is stored.

For each ego belief, write its mind energy counterpart on the right side.

Not the forced positive version.

Not a fake affirmation the body rejects immediately.

The deeper, quieter truth beneath the noise.

If the archive says, I am not enough, the truth is not necessarily I am enough shouted louder. The deeper truth is: enough is not the relevant metric. I am.

If the archive says, I must prove myself constantly or I will be seen as worthless, the truth is: existence does not require proof. Proof belongs to performance. Being precedes performance.

If the archive says, I cannot trust anyone, the truth is: my body learned protection, but protection is not the same as truth.

If the archive says, I am behind, the truth is: life is not a single race with one clock.

If the archive says, I must control everything to be safe, the truth is: control is not safety. Presence is closer.

Write slowly.

Feel the body's reaction to each sentence.

The true sentence will not always feel comfortable. Sometimes truth irritates the archive at first because the archive has been protecting its old files like sacred law. But beneath the irritation, there is often a small loosening.

A breath.

A drop in the shoulders.

A slight opening behind the sternum.

That is the signal.

This is not an exercise in reprogramming the ego with better slogans. It is an exercise in discovering that there is something behind the ego that was never programmed by anyone. Something that has never needed to prove anything. Something that is reading these words right now with a quiet the archive cannot quite reach.

That quiet is mind energy.

It has always been there.

You are not creating it.

You are noticing it.

CHAPTER SIX

THE ENERGY VORTEX
Chakras, Crystals, and the Human Engine

"When we align our energy centers, we do not become powerful. We remember that we always were."

THE CLUSTERED NODES

Close your eyes and bring your attention to the center of your chest.

Do not visualize anything yet.

Just feel.

There is pressure there. Warmth. Movement. The ribs rise and fall. The heart contracts behind bone and muscle. Breath passes through the throat above it. The stomach moves beneath it. The body is not a blank space with organs randomly placed inside. It has centers. Regions. Gatherings. Places where sensation collects more intensely than others.

You know this already.

Fear drops into the gut.

Grief presses into the chest.

Shame burns in the face.

Truth gets stuck in the throat.

Desire gathers low in the belly.

Intuition tightens behind the eyes.

This is not superstition.

This is embodied mapping.

Let me tell you what I think happened when the ancient seers of India developed the chakra system.

They had a practice. They sat still. Not casually. Not for three distracted minutes between appointments. They sat with sustained attention turned inward for periods of time that would seem impossible to a modern nervous system trained by screens, alerts, urgency, comparison, and noise.

They refined this practice across generations.

Teacher to student.

Body to body.

Breath to breath.

Each generation added precision. Each practitioner became a kind of living laboratory. No microscope. No MRI. No hormone panel. No electrocardiogram. Just the human organism turned inward with ruthless patience.

And in that disciplined inner science, they observed something.

They observed that there are specific locations in the human body where awareness naturally concentrates. Places where sensation thickens. Places where emotion gathers. Places where experience organizes itself around recurring themes: survival, sexuality, power, love, expression, insight, connection.

They called these concentrations chakras — wheels, turning centers, vortices of prana, life-force.

A wheel is not random symbolism.

A wheel turns around a center.

A vortex draws movement into pattern.

A node gathers signal.

That is what they felt.

They mapped seven primary centers: at the base of the spine, just below the navel, at the solar plexus, at the center of the chest, at the throat, between the eyebrows, and at the crown of the head.

Seven locations.

Seven themes.

Seven ways the human being meets reality.

For centuries, Western medicine looked at this map and said: there is nothing anatomically at those locations that corresponds exactly to a spinning wheel of energy.

And Western medicine was right in the narrow sense.

You cannot cut open the body and pull out a chakra like an organ. You cannot weigh it on a scale. You cannot point to one single structure and say: there, that is the whole thing.

But that was the wrong question.

A city is not only one building. A network hub is not only one cable. An airport is not one runway. A chakra is not one anatomical object. It is a functional region — a cluster of nerves, glands, vessels, fascia, sensation, memory, and meaning. A place where body, emotion, and awareness meet.

The ancient map was not crude anatomy.

It was functional cartography.

It mapped experience.

And when you place that experiential map beside what modern biology knows about nerve plexuses, endocrine glands, breath, vagal pathways, interoception, posture, and social regulation, the overlap becomes difficult to ignore.

Not identical.

Meaningful.

The root chakra corresponds to the region of the pelvic floor, legs, base of the spine, lower body, and the survival systems connected to grounding, elimination, movement, reproduction, and threat response. The adrenal glands are not located at the base of the spine, but their survival chemistry belongs to this same theme: fight, flight, freeze, mobilize, endure.

When the root is described as safety and survival, that is not mythology.

That is the body asking: Am I safe enough to be here?

You can feel this.

When you feel unsafe, the body does not become philosophical. The pelvic floor tightens. The legs brace. The breath shortens. The gut changes. The eyes scan. The system prepares for impact or escape.

Safety is not an idea.

Safety is muscle tone, breath depth, grounded feet, regulated chemistry, and the body's permission to stop bracing.

The sacral chakra corresponds to the lower abdomen, reproductive organs, pelvic bowl, hips, and the deep waters of sensation, sexuality, creativity, pleasure, and emotional flow. This is the region of movement and generation. The belly softens or locks. The hips open or guard. The body either allows life to move through or clamps down around it.

The enteric nervous system contains over 500 million neurons and produces a significant portion of the body's serotonin. While most of this serotonin operates locally in the gut rather than directly shaping mood, the system communicates continuously with the brain through the vagus nerve, influencing emotional tone, stress response, and perception.

The belly knows.

It does not know in sentences.

It knows in tightening, warmth, nausea, hunger, ease, contraction, appetite, disgust, pull, and release.

When the sacral center is described as creativity and sensuality, that is not decorative spirituality. Creativity is not only a mind event. It is generative force. It is the body's capacity to bring something into form. A child. A song. A business. A meal. A room. A sentence. A life.

Creation begins below language.

The solar plexus corresponds to the upper abdomen, diaphragm, stomach, liver region, and the felt center of personal agency. This is where will becomes heat. Where decision becomes posture. Where the body says yes, no, go, stop, stand, act.

You know the solar plexus by impact.

A hard truth can hit you there.

A betrayal can knock the air out of you.

A surge of courage can rise from there like fire.

This is the body's command center for power as lived experience. Not domination. Agency. The felt sense that you can act in the world and your action matters.

When this center collapses, the body folds inward.

When it is inflated, the body pushes outward aggressively.

When it is coherent, the spine lengthens and the breath has authority.

Power is not volume.

Power is organized will.

The heart chakra corresponds to the chest, lungs, heart, breath, circulation, arms, hands, and the social body's capacity for connection. The heart is electrically active. Every beat begins as signal. The lungs bring outside air into inside blood. The arms reach outward. The hands touch. This region is where separation becomes contact.

When grief comes, it lands here.

When love comes, it opens here.

When betrayal comes, it armors here.

The heart is not just a symbol because poets got lazy. The heart became the symbol because the body kept pointing there.

Love changes the chest.

Loss changes the chest.

Prayer changes the chest.

Forgiveness changes the chest.

The traditions were not inventing symbolism out of thin air. They were naming the body's own geography.

The throat chakra corresponds to the throat, jaw, tongue, larynx, vocal cords, breath channel, and the nerves and muscles involved in speech, swallowing, expression, and social engagement. Voice is not only sound. Voice is identity leaving the body through vibration.

To speak, air must rise from the lungs, pass through the vocal cords, vibrate, and be shaped by mouth, tongue, teeth, and lips into language.

Thought becomes breath.

Breath becomes vibration.

Vibration becomes word.

Word changes the world.

That is not small.

When truth is withheld, the throat knows. It tightens. Swallowing changes. The jaw locks. The voice becomes thin or forced. When truth moves cleanly, the throat opens. Sound becomes full. Words carry weight without needing to shout.

Authentic expression is not saying everything.

It is saying what is true enough for the body not to split itself in half.

The third eye corresponds to the brow, visual system, frontal attention, imagination, insight, and the capacity to perceive pattern beyond immediate surface. Close your eyes and picture a place you love. You are not seeing it with the eyes, but vision is happening. The inner screen lights up. The mind forms image. Memory, perception, and possibility meet.

This center is not fantasy.

It is pattern recognition.

Insight is the moment scattered data becomes shape.

The mechanic hears the engine and knows the problem. The mother hears a cry and knows which child it is. The surfer sees the ocean standing up far outside and knows where the wave will break. The artist looks at the canvas and sees the next line before the hand moves.

That is third-eye function in practical clothes.

Perception beyond the obvious.

The crown corresponds to the top of the head, the felt vertical axis of the body, and the human capacity to experience connection with the larger field of existence: God, source, consciousness, cosmos, silence, meaning, the whole. Different traditions name it differently because language fractures at altitude.

The body often feels this as openness upward, lightness,

expansion, stillness, reverence, or the strange sensation that the top of the head is not a ceiling but a doorway.

That experience is not owned by any one tradition.

People feel it in churches, temples, forests, births, deathbeds, oceans, mountains, meditation, music, and moments when the self briefly stops making itself the center of the universe.

Two knowledge systems.

Different tools.

Different centuries.

Different languages.

One inner observation refined through disciplined attention. One outer observation refined through instrumentation and anatomy. Both describing the same human instrument from different sides of the glass.

This is the point.

The chakra system is not a cartoon of colored wheels floating in the body.

It is an ancient map of clustered human experience.

A map of where life becomes fear, pleasure, power, love, truth, vision, and union.

ALIGNMENT AS ENGINE

When these centers are aligned — not merely active, but coherent with one another — something qualitatively different happens in the system.

Think mechanically.

A single cylinder firing in an engine produces movement, but also vibration. Multiple cylinders firing randomly produce noise, heat, inefficiency, and damage. Multiple cylinders firing in precise sequence produce power. The fuel is the same. The metal is the same. The difference is timing.

Alignment converts availability into function.

The human engine works the same way.

You can have survival active without love, and the result is defense.

You can have desire active without truth, and the result is addiction.

You can have power active without heart, and the result is control.

You can have vision active without grounding, and the result is fantasy.

You can have spirituality active without the lower body, and the result is escape wearing sacred language.

This is where people get lost.

They try to open the crown while abandoning the root. They chase light while ignoring their body. They speak of universal love while their nervous system is in threat. They perform peace while their jaw is locked and their gut is burning.

That is not alignment.

That is bypassing.

The engine does not run clean because one cylinder is loud.

It runs clean when all cylinders fire in sequence.

Root gives safety.

Sacral gives flow.

Solar plexus gives agency.

Heart gives connection.

Throat gives expression.

Third eye gives perception.

Crown gives orientation to the whole.

Each center needs the others.

A house teaches this. Foundation without roof is not shelter. Walls without windows are confinement. Plumbing without drainage is disaster. Electrical wiring without grounding is danger. A beautiful room on a broken foundation will crack. The whole structure matters.

So does the body.

The root says: I am safe enough to be here.

The sacral says: life can move through me.

The solar plexus says: I can act.

The heart says: I can connect.

The throat says: I can express.

The third eye says: I can see.

The crown says: I belong to the whole.

When these statements contradict each other, the body fragments.

When they support each other, coherence arrives.

You have felt this.

A moment when your feet were grounded, your breath was deep, your heart was open, your voice was clear, your mind was sharp, and your actions matched your values. No internal argument. No static. No performance. Just clean signal.

That is alignment.

Not fireworks.

Function.

This is the experience traditions call awakening. Not necessarily the sudden supernatural detonation people imagine. The daily, available awakening of a system coming into right relationship with itself. The feeling of something clicking into place. Static clearing. Signal emerging. The body no longer fighting its own instruction.

Awakening is not becoming special.

It is becoming coherent.

A musical instrument proves this. A guitar can have all six strings present. But if each string is tuned to the wrong tension, every chord hurts. Nothing is missing. The instrument is complete. It is simply not tuned.

Tuning does not add strings.

It restores relation.

The practice of alignment is not a one-time achievement. It is daily tuning. Use, stress, grief, joy, sleep, food, conflict, weather, family, money, work, screens, memory — all of it shifts the instrument. This is not failure. This is life touching the strings.

You tune because you are alive.

QUANTUM ENTANGLEMENT AND SELECTIVE VIBRATION

Here is where the physics becomes remarkable, and where precision matters.

We have established that the body is a living electrical instrument. We have mapped clustered nodes of sensation, nerve activity, endocrine function, breath, memory, and meaning. Now consider the deeper structure of connection itself.

Quantum entanglement shows that, at the smallest scales, particles that have interacted can remain correlated in ways that defy ordinary intuition about distance. Measure one part of an entangled system, and the total system displays correlations that cannot be explained by classical separation alone. Einstein disliked this deeply and called it "spooky action at a distance." Later experiments, including increasingly refined Bell test experiments, confirmed that nature is stranger than classical common sense.

This does not mean human emotions are quantum signals; it means reality is relational at its deepest measurable scales.

Do not cheapen the mystery by making it sloppy.

It means reality at its root is relational in ways the ordinary mind cannot fully picture.

Separation is not as absolute as it appears.

That is the doorway.

The atoms in your body were forged through cosmic processes: hydrogen from the early universe, heavier elements formed in stars and stellar explosions, matter circulating through time until it became ocean, soil, food, blood, bone, nerve, thought. Every atom in you has a history older than your name. Every breath participates in a planetary exchange. Every cell is made from matter that has been elsewhere, in other forms, through other bodies, under other suns.

You are not isolated material.

You are continuity with skin.

That is enough to say with force.

Connection does not require exaggeration. It is already staggering.

You are connected to the cosmos through origin, through matter, through fields, through light, through gravity, through breath, through ancestry, through biology, through every exchange that allows this body to continue another day.

But connection is not the same as resonance.

This is crucial.

A radio antenna is surrounded by countless signals. It does not play all of them. It plays the frequency it is tuned to receive.

The room is full of transmissions — music, emergency channels, aircraft communication, cellular data, satellite signals — but the receiver must be tuned, powered, and configured.

Your body is similar.

You are immersed in fields, signals, people, environments, histories, pressures, memories, and possibilities. You cannot consciously process all of them. You resonate with what your system is tuned to notice, expect, fear, desire, and embody.

A fearful system finds threat quickly.

A resentful system finds insult quickly.

A grateful system finds support quickly.

A creative system finds pattern quickly.

This is not because the world magically changes to flatter your mood.

It is because attention selects.

The instrument amplifies what it is tuned to receive.

A room full of people proves it. One person enters looking for rejection and finds it in a delayed glance. Another enters looking for connection and finds it in the same room through one warm voice. Same room. Different tuning. Different reality experienced.

This is why inner alignment matters.

Not because the outer world is imaginary.

Because your state determines which parts of the outer world become available to you.

Why do we forget that power comes from the center within, and yet willingly submit to frequencies outside ourselves?

The answer is simple and uncomfortable.

Ego is louder than stillness.

Noise is easier than silence.

Comparison gives the ego measurement data, and alignment

gives the ego nothing to brag about. The ego wants scoreboard, audience, proof, reaction. Stillness offers no applause. Alignment does not flatter the personality. It simply returns the system to proper function.

That is why the ego resists it.

The center is quiet.

The center does not market itself.

The center does not compete.

It waits.

The engine is always available. The nodes are always capable of alignment. The body is always speaking. The breath is always offering a doorway back. The ground is always beneath you. The heart is always pulsing. The throat is always holding what has not been said. The belly is always reporting what the mind tries to override.

You need only bring attention back within.

Only.

That word is funny because it is the simplest instruction and the hardest human task.

Bring attention back.

Again.

Again.

Again.

Not once.

As a life.

LIVING INTEGRATION: THE SEVEN-POINT SCAN

Lie down comfortably.

Let your body be fully held by whatever is beneath you. Floor, bed, mat, earth. Feel the contact. Back of the head. Shoulder blades. Spine. Pelvis. Legs. Heels. Let gravity do what it has been doing your entire life.

Close your eyes.

Do not try to become spiritual.

Become accurate.

Beginning at the base of your spine, bring your attention — not your hand, not a forced visualization, just bare attention — to each center. Spend at least sixty seconds at each location. The quality of attention matters more than technique.

Attention is light.

Where you place it, something becomes visible.

At the base of the spine, the root:

Do I feel safe?

Not should I feel safe.

Not am I safe according to logic.

Do I feel safe in the body right now?

Feel the pelvic floor, legs, feet, and the points of contact with the surface beneath you. Notice bracing, numbness, heaviness, restlessness, ease. Your relationship with the ground is not abstract. It lives in muscle tone.

Let the body register: I am held.

Just below the navel, the sacral center:

Do I feel life moving?

Feel the lower belly, hips, pelvis. Is there warmth, fluidity, softness, vitality? Or numbness, gripping, shame, tightness, absence? Do not judge what you find. Judging is just another contraction.

This is the water center.

Let it be water before you demand it become a river.

At the solar plexus:

Do I feel my will?

Bring attention to the upper abdomen, diaphragm, stomach, and the region beneath the ribs. Notice heat, pressure, collapse, agitation, strength. This is where agency becomes physical. This is where yes and no begin to gather force.

Ask: Where in my life have I abandoned my own action?

Then breathe.

Not to fix.

To listen.

At the center of the chest, the heart:

Do I feel love?

Not the idea of love.

The felt sense.

Warmth, ache, openness, grief, pressure, guardedness, tenderness. The heart center may not feel soft at first. Sometimes it feels armored because it has had reasons to protect itself. Respect that. Armor is not failure. Armor is history.

Place attention there like a hand on a locked door.

No force.

Presence.

At the throat:

Do I feel free to speak?

Notice the jaw, tongue, throat, neck, and breath. Is there tightness? A swallowed sentence? A truth waiting behind the teeth? A fatigue from saying what is acceptable instead of what is accurate?

The throat is where the invisible becomes audible.

Let it know you are listening.

Between the eyebrows, the third eye:

Do I feel clear?

Bring attention to the brow, eyes, forehead, and inner screen of the mind. Notice whether thought is sharp, scattered, foggy, pressured, or calm. Clarity is not more thinking. Clarity is the moment thought stops blocking perception.

Ask: What am I seeing that I keep pretending not to see?

Then be still.

At the crown:

Do I feel connected?

Not to a concept. Not to a doctrine. To the larger field. To life beyond the personal story. To God, source, universe, silence, consciousness — whatever word opens the top of the body instead of closing the mind around an argument.

Feel the space above the head.

Feel the vertical line of the body.

Feel that you are not separate from the whole that is holding you.

For any center where the answer is faint or absent, breathe into it. Not forcing it open. Not imposing a state that is not there. Just offering attention and waiting to see what responds.

The engine tunes itself when attention is honest.

The intelligence of the body knows more than the ego thinks it knows.

Your job is to show up.

To be present in each location.

To let the system speak.

Alignment is not becoming powerful.

Alignment is removing the interference from power that was already there.

CHAPTER SEVEN

SACRED GEOGRAPHY
The Standing Waves of Place

"There are places on this earth where the veil is thinner. Where the peak never subsides."

WHY PLACE MATTERS

You have felt it.

Walk into a room and your body knows before your mind has evidence.

The air changes. Not chemically in any obvious way you can name, but physically in the way your chest responds. Breath shortens. Skin tightens. The eyes scan the corners. The stomach drops half an inch. Something in the space feels dense, wrong, charged, as if the room is still holding a sound after the sound has ended.

Then later you learn there was an argument there. A violent one. Voices raised. Bodies tense. Someone leaving. Someone staying. The room had no mouth, but it told you.

You have felt the opposite too.

You step onto land and the body rises. The breath widens.

The shoulders drop. The field of vision opens. You feel lighter, not as an idea, but in posture. The soles of the feet meet the ground differently. The chest lifts. The mind becomes less crowded. And later you learn that people have come to that place for generations to pray, to bury, to begin, to ask forgiveness, to receive instruction from the silence.

The body knew.

Before the brochure.

Before the history.

Before the mind had a reason.

This is why place matters.

Place is not backdrop.

Place is an active participant.

The modern mind thinks of location as coordinates: address, map pin, parcel, elevation, zoning, distance from airport, distance from ocean, distance from work. Useful information. Necessary information. But it is not the whole truth of a place.

A place has pressure.

A place has tone.

A place has memory.

A place has geometry.

A place has light behavior, sound behavior, wind behavior, water behavior, mineral behavior, and human behavior layered into it through time.

You do not enter a place as a neutral observer. You enter it as an instrument. Your body begins reading immediately. Temperature. Humidity. Air movement. Light angle. Reverberation. Smell. Magnetism. Altitude. Openness. Threat. Safety. History. Human residue. The nervous system gathers all of it faster than the conscious mind can turn any of it into explanation.

This is not superstition.

This is perception operating below language.

Your liquid crystal body is registering real information about the environment: the acoustic quality of the room, the mineral composition of the ground, the movement of air, the presence of water, the angle of sunlight, the emotional state of other bodies, the accumulated patterns of human attention, the way the space amplifies or deadens vibration.

A room with low ceilings and no airflow feels different from a cliff edge facing the sea.

A hospital hallway feels different from a chapel.

A battlefield feels different from a garden.

A childhood home feels different from a hotel room, even if both have walls, beds, windows, and doors.

Why?

Because place stores relationship.

Not as a ghost story.

As pattern.

A wooden floor records footsteps as wear. A stone path records centuries of walking as polish. A room records smoke in fabric, water damage in corners, sound in the way it echoes, heat in the walls, human behavior in arrangement. Chairs facing each other invite conversation. Chairs facing a screen invite consumption. A doorway frames entry. A window frames longing. Architecture instructs the body.

Place tells you what to do with yourself.

Certain places multiply that instruction.

They become more than locations.

They become fields.

Certain places have qualities that are intensified by the

convergence of multiple forces: light, sound, water, wind, geology, human attention, ancestral memory, architecture, and the particular geometry of the land. When these forces align at a location, the effect is not merely additive. It becomes patterned. Coherent. Strong enough to alter the body's state.

Physics has a name for a stable pattern of wave activity that holds position.

A standing wave.

A standing wave is a wave pattern that does not appear to travel through space in the ordinary way. It holds. It forms fixed nodes, where motion is minimal, and fixed antinodes, where amplitude is greatest. A guitar string vibrating between two fixed points creates a standing wave. A flute creates standing waves inside a column of air. A room can hold standing waves of sound depending on its shape. A tide pool can hold repeated wave behavior. A canyon can catch echo. A cathedral can hold chant.

The wave is not just passing through.

The place is shaping it.

The geometry holds the pattern.

Some places on Earth are like this.

They are antinodes.

They hold peak experience with unusual consistency. Not because anyone invented the feeling. Not because culture imagined it into existence from nothing. Culture often arrives afterward. Culture names what the body was already feeling.

A sacred place becomes sacred because something there already concentrates attention.

Then human attention returns to it again and again, amplifying the pattern further.

A bell rings differently depending on the metal, the shape, the strike, and the room. A place is the same. Geology is the metal. Geometry is the shape. Weather is the strike. Human attention is the ringing.

Some places ring for centuries.

THE REFRACTIVE WAVE ISLAND

I live in the middle of the Pacific Ocean.

That sentence should be felt in the body.

Not as a travel description. As a physical condition.

An island in the middle of the largest ocean on Earth is not merely land surrounded by water. It is a body held inside motion. Waves arrive from distances so vast the mind cannot hold them cleanly. Wind crosses open water and reaches land carrying salt, temperature, pressure, and information. Weather forms, breaks, bends, and reforms. Light reflects from water in all directions. Clouds gather around mountains. Rain falls on ridges. Streams cut valleys. Valleys return water to the sea.

The island is not sitting there.

The island is interacting.

The Pacific does not surround it passively. It speaks into it from every side.

A hotspot erupts.

A new island emerges.

The old island remembers.

That age matters.

You can feel it.

The rock has been worked on. Not by one force, but by many. Rain has entered it. Rivers have cut it. Wind has carved it. Ocean has struck it. Roots have penetrated it. Minerals have moved through it. Red dirt has formed from iron-rich volcanic material weathered under tropical conditions. The land is not raw fire anymore. It is fire softened by water and time.

That creates a different tone.

A younger volcanic landscape can feel exposed, active, primal, unfinished, while another island feels ancient in the body. Green over red. Water over fire. Birdsong over stone. Erosion not as loss, but as revelation. The island has been opened by time.

The geometry is extraordinary.

A central high place. Ridges radiating outward. Valleys cut like channels. Waterfalls dropping through green walls. Coastlines shaped by surf, cliff, reef, sand, lava, and collapse. Wind moving over ridges and funneling through valleys. Rain falling heavy in one region while another sits in gold light. The island behaves like an instrument whose chambers are valleys, whose strings are ridgelines, whose breath is wind, whose drum is the sea.

The best phrase I have found for it is this:

A refractive wave peak simulation.

Not simulation as fake.

Simulation as a living model.

Energy arrives at the island from every direction: ocean swell, wind, sunlight, lunar pull, atmospheric pressure, human attention, geological memory. The island's shape bends it. Ridges redirect wind. Valleys channel sound. Cliffs amplify surf. Waterfalls atomize water into air. Forest canopy filters light. Red dirt absorbs heat. The ocean reflects light back upward. Reef turns wave into thunder. Everything bends, breaks, reflects, and returns.

Refraction is the bending of a wave when it enters a different medium or encounters changing conditions. Light bends through glass. Waves bend around reef. Sound bends through air shaped by temperature and terrain.

A sacred place bends energy.

That is the sensation.

Nothing arrives straight and leaves unchanged.

The shoreline rises from the ocean in fluted cliffs, deeply grooved by waterfalls and time. Those valleys do not merely look dramatic. They function like resonant chambers. Wind moves through them. Surf strikes below. Water drops from heights. Sound folds and returns. The body receives it before the mind identifies it.

Stand there and the chest changes.

The ordinary word "wind" becomes too small.

This is not just air moving.

It is air carrying the shape of the valley.

The people of this land understood wind with specificity because wind was not generic. Wind had names, directions, moods, behaviors, meanings. A wind coming through one valley is not the same as wind coming off open water. A misting rain is not the same as hard rain. A cloud on a ridge is information. A change in bird behavior is information. The environment is not mute if a culture remains fluent enough to read it.

This is what modern life has damaged.

Not intelligence.

Fluency.

Stand at the edge of a canyon and look into the red. The body registers depth. Distance. Time. Iron. Exposure. The walls hold color like blood dried into stone. Light moves across the canyon and reveals scale beyond ordinary human proportion. You become smaller, but not diminished. Right-sized.

That is one of the gifts of sacred geography.

It restores proportion.

The iron-rich red of the basalt walls is not only visual. It carries elemental kinship. The iron in those walls and the iron in your blood belong to the same larger story of stars, minerals, pressure, and time. To stand at the canyon's edge is to stand beside geological memory and feel the body recognize its own ingredients outside itself.

The mountain is not other.

The canyon is not other.

The blood and the wall are cousins.

It is not property first. It is connection to source.

That distinction changes everything.

Care comes first.

Reciprocity comes first.

You do not simply take from what feeds you. You enter into obligation with it.

THE EIGHT ENERGIES OF PLACE

Through years of being attentive to the felt quality of places — across the islands of Hawaiʻi, in places on the continent where I have spent time, and in other parts of the world where the body immediately recognized something unusual — I have come to recognize eight primary types of place-energy.

Call them forces.

Call them inputs.

Call them signatures.

Each contributes to the electromagnetic, sensory, emotional, and symbolic character of a location.

The first is light and frequency.

Every place has its own light behavior. Not simply bright or dark. The angle, color, reflectivity, intensity, filtering, and timing of light. Morning light entering through trees is not the same as afternoon light off concrete. Desert light is not forest light. Ocean light is not mountain light. Light filtered through stained glass is not light reflected from snow.

The body responds differently to each.

Light affects hormones, circadian rhythm, mood, alertness, and perception. A room with no natural light changes the nervous system. A coastline at sunrise changes it differently. A forest canopy creates a green-diffused light that softens the eyes. A lava field in full sun creates glare, heat, exposure, and vigilance. A temple lit by candles invites inward attention because the light is small, warm, and localized.

Light tells the body what kind of world it is in.

The second is sound and vibration.

Every place has a base tone. Not always audible as a single note, but present as the ongoing sound beneath events. Surf. Wind. Insects. Traffic. Water. Electrical hum. Silence so deep it becomes a sound of its own. Reverberation from walls. Echo from stone. Absorption by trees.

Sound enters the body physically. Pressure waves strike the eardrum, move tiny bones, bend hair cells, become signal. Low frequencies are felt in the chest and gut. Repetitive rhythms entrain breath and attention. Chant in a stone chamber changes the body differently than speech in a carpeted office.

Ancient ceremonial sites understood this. So do churches. So do concert halls. So do caves. So do stadiums. A crowd chanting in unison is not just making noise. It is synchronizing bodies.

Sound organizes groups.

Sound organizes breath.

Sound organizes emotion.

The third is ancestral connection.

Some places carry the weight of human continuity. Birth, death, prayer, labor, feast, suffering, ceremony, song, burial, return. The mind may not know the details, but the body can feel age. A place used by humans for generations often has a density that a new commercial space does not.

This is hard to reduce to one mechanism, and it does not need to be reduced to one mechanism. Memory lives in material traces, stories, architecture, pathways, names, ritual, bones, objects, and repeated human behavior. The place becomes a container of continuity.

An old church feels different from a new event room.

A family kitchen feels different from a model home.

A spiritual platform feels different from a parking lot.

The difference is not only belief.

It is accumulated attention.

A place remembers how it has been treated.

The fourth is spiritual consciousness.

This is the accumulated effect of intentional attention over time. Prayer, meditation, chanting, ceremony, grief, confession, devotion, healing, gratitude. When many people bring their most concentrated attention to the same place for years, decades, centuries, the place changes in how it receives attention.

A path becomes easier to walk when it has been walked.

A chant becomes easier to enter when it has been chanted.

A room becomes easier to pray in when prayer has soaked into its use.

This is not because the walls are magical in a childish sense. It is because human beings are patterning creatures. Places shape practice, and practice shapes places. The cushion in the corner teaches the body to sit. The altar teaches the eyes to soften. The chapel teaches the voice to lower. The trail teaches the feet where to go.

Over time, the place becomes a carrier wave for the practice.

That is why pilgrimage works. The traveler is not only going somewhere symbolic. The traveler is entering a field of accumulated intention. Mecca, Jerusalem, Varanasi, Santiago de Compostela, Haleakalā, Mauna Kea, Wailua, the many heiau of Hawaiʻi — sacred places are not interchangeable. They have been shaped by attention and returned to attention for generations.

The fifth is potential and motion.

Moving water carries a different charge than still water. A waterfall changes air. Surf changes air. A river confluence changes sound, humidity, ions, movement, and mood. Wind moving through trees changes the nervous system differently than stagnant indoor air. Tectonic tension, volcanic activity, seasonal storms, migrating birds, blooming cycles, tides — all create the sense that the place is not fixed.

Motion wakes the body.

A waterfall does this immediately. You feel mist on the skin. You hear the continuous impact. Negative ions may contribute to the effect, but the experience is larger than one measurable factor. The body understands falling water as power, cleansing, danger, renewal, and movement all at once.

Surf does the same.

A breaking wave is stored distance arriving as impact.

Storm energy traveled across ocean, organized into swell, rose over reef, and collapsed into sound, foam, and force. Stand near heavy surf and the ribs know it. The body feels the percussion.

The sixth is space and ether.

Openness matters.

A wide view changes breathing. Stand on a cliff, mountain, open beach, or high ridge, and the lungs often expand without instruction. The eyes travel farther. The mind loosens its grip. Enclosed spaces can focus attention, but they can also constrict it. Open spaces allow the nervous system to orient across distance.

The ratio of sky to ground matters.

The movement of air matters.

The horizon matters.

A horizon gives the mind a line between here and beyond. It lets the body locate itself inside vastness without being swallowed by it. This is why ocean horizons calm some people. The line is clean. The movement is continuous. The scale is large enough to absorb the small noise of the self.

Space is medicine when the mind has become crowded.

The seventh is atomic and material.

The ground beneath you matters. Basalt, granite, limestone, coral, clay, sand, iron-rich soil, quartz-bearing rock — different materials hold heat differently, conduct differently, reflect differently, erode differently, sound differently underfoot. The body receives these differences even when the mind does not name them.

Walk barefoot on black lava rock warmed by sun.

Then walk barefoot on wet grass.

Then on cold tile.

Then on sand.

Different world.

Different body.

Material is not neutral.

Iron-rich volcanic soil carries one feeling. Coral sand carries another. Forest loam carries another. Granite mountain carries another. The minerals in water change taste. The mineral structure of stone changes resonance. The conductivity of wet ground changes electrical relationship. The land's chemistry enters plants, food, water, and body.

You are not standing on scenery.

You are standing on composition.

The eighth is elemental creation.

Earth, water, fire, air.

The old elemental language survives because it maps experience cleanly. Earth is solidity, weight, form, mineral, ground. Water is flow, dissolution, reflection, emotion, circulation. Fire is heat, ignition, metabolism, sunlight, volcano, digestion, will. Air is breath, movement, weather, voice, exchange.

Volcanic islands concentrate these elements intensely.

Earth rising from ocean.

Fire making land.

Water carving it.

Air shaping weather.

Life covering the whole thing in green.

In geothermal areas, this is not abstract. It is visible. Lava becomes rock. Rock becomes soil. Rain becomes stream. Stream becomes valley. Valley becomes loʻi, forest, waterfall, river mouth, reef. Fire feeds earth. Earth feeds water. Water feeds life. Air moves all of it.

Elemental creation is not mythology.

It is process.

When multiple energy types peak simultaneously at one location, the effect becomes multiplicative. Light amplifies sound.

Sound amplifies ancestral memory. Water amplifies motion. Geology amplifies field. Human attention amplifies place. Place amplifies attention.

That is why certain places feel overwhelming in the best sense.

More reality per breath.

More signal in the body.

More silence in the mind.

These places are energetic antinodes.

They hold the peak.

LIVING INTEGRATION: SENSING THE STANDING WAVE

The next time you visit a place that moves you, stop before you do anything else.

Do not take the photo first.

Do not narrate it.

Do not make yourself the center of the scene.

Stop.

Feel your feet. Let the body register where it has arrived. Notice the air on your skin. Notice whether the breath rises or drops. Notice whether the jaw tightens or releases. Notice the first emotional color before the mind starts explaining it.

Ask: what is present here?

Light.

How does it fall? Direct, filtered, reflected, harsh, soft, gold, blue, green, dim? Where does it land on your body?

Sound.

What is the base tone? Water, wind, bird, silence, machinery, echo, human voice? Where do you feel it — ears, chest, bones, stomach?

Ancestral connection.

Does the place feel old? Watched? Held? Heavy? Welcoming? Solemn? Does the body lower its voice without being asked?

Spiritual consciousness.

Do your thoughts become quieter? Cleaner? More honest? Does the place invite prayer, even if you do not use that word?

Potential and motion.

What is moving? Water, wind, cloud, animal, leaf, tide, heat, people, shadow? Does the movement wake you up or calm you down?

Space.

Can your eyes travel? Does the view open the lungs or close them? Is there room for the mind to spread out?

Material.

What is under you? Stone, soil, sand, wood, concrete, grass? What temperature does it hold? What does it ask of your feet?

Element.

Which element dominates? Earth, water, fire, air? Which one is missing? Which one is calling your attention?

Do not turn this into analysis too quickly.

Let the body report.

Notice what shifts in you. What becomes available that was not available before you arrived? What quiets? What rises? What memory appears? What decision becomes clear? What grief softens? What part of you stands up straighter?

The standing wave is not something you observe from outside.

It moves through you.

You are the instrument it plays when you arrive.

Your job is not to own the place.

Your job is to listen well enough to know what music you have become.

CHAPTER EIGHT

THE SIMULATION AND THE SONG

"We create stories within the story that created us."

THE MACHINE WITHIN THE MACHINE

Look at a screen.

Not the content.

The screen itself.

Glass, light, pixels, charge, code. A flat surface pretending to be depth. Tiny points of illumination arranged so quickly and precisely that your eyes stop seeing the machinery and begin seeing a world. A face. A landscape. A war. A child laughing. A message from someone you love. A movie. A memory. A map.

The screen is not the thing it shows.

And yet your body responds.

Your heart can race from a scene that is not physically happening. Your eyes can fill with tears from a face made of light. Your nervous system can tighten from words typed by a person miles away. The body does not wait for philosophy to decide whether the experience is real enough.

It reacts to pattern.

That should tell us something.

Here is a thought that will either disturb you or liberate you, depending on your relationship with control, and possibly do both at once, which is the signature of ideas with real voltage:

You may be living inside a simulation.

Before you close this book or dismiss the idea as the sort of thing people say at two in the morning after too much philosophy and not enough sleep, let me be precise about what I mean — and more importantly, what I do not mean.

I do not mean the childish nihilistic version: nothing is real, nothing matters, we are all just pixels in some cosmic video game, so ethics is optional and meaning is a delusion.

That is not liberation.

That is avoidance dressed as insight.

If nothing is real, then suffering is not real. If suffering is not real, compassion becomes optional. If compassion becomes optional, cruelty can call itself intelligence. That road does not lead to awakening. It leads to vacancy.

A person who uses metaphysics to escape responsibility has not become profound.

They have become slippery.

That is not what I mean.

What I mean is something far more beautiful and far more serious.

The universe appears to behave computationally. Not in the small, dead sense that reduces reality to machinery, but in the deeper sense that reality processes information according to consistent rules. Mathematical rules. Reliable rules. Rules that hold whether or not anyone believes in them. Rules that let atoms form, stars burn, planets orbit, cells divide, hearts beat, and light travel through space at a defined limit.

The laws of physics behave like executable structure.

Not metaphorically only.

Operationally.

Input conditions produce patterned outcomes. Change the conditions, and the output changes according to relation, constraint, and law. Gravity curves motion. Electromagnetism governs charge. Quantum fields produce probabilities. Thermodynamics sets direction. Geometry influences the path. Information is conserved, transformed, distributed.

Reality is not random noise.

Reality runs.

A computer takes symbol, rule, voltage, and time, then generates image, sound, movement, and simulated world.

The universe takes matter, law, light, and time, then generates stars, oceans, bodies, minds, grief, hunger, music, memory, and the strange ache of asking why.

The machine we built resembles the machine that built us.

That is not an accident.

It is recursion.

If the universe is a computation, what is it computing?

Experience.

That is the answer that keeps returning.

It is computing the conditions under which consciousness can encounter itself. It is producing the contrast required for knowing: self and world, light and shadow, memory and anticipation, longing and arrival, birth and loss, observer and observed.

A simulation is not necessarily fake.

A simulation is a structured environment where experience can occur under rule.

Flight simulators are not useless because the pilot is not actually in the sky. They teach the body how to respond. Architectural models are not meaningless because no one lives inside them. They reveal structure before construction. A musical score is not the song, but it allows the song to appear when breath, hand, instrument, and time enter it.

The simulation is not the opposite of reality.

Simulation is one of reality's methods.

You are a process in that computation while also being a spectator of it. You are inside the song while hearing it. You are a character in the story while also made of the language that writes the story. You are not outside creation looking at a machine. You are creation running locally as a body, a mind, a field of attention.

Every sensation you experience is data.

Every emotion is measurement.

Every choice is input.

Every consequence is output.

Every moment of awareness is the computation doing what it was designed to do: experience itself from within.

Meaning is not somewhere else.

That is the mistake.

Meaning is not waiting at the end of a secret path, hidden behind the right career, the right relationship, the right retreat, the right achievement, the right doctrine. Meaning is not a prize buried under enough spiritual effort.

The meaning is that experience is happening.

Here.

Through you.

The universe spent billions of years arranging hydrogen, gravity, heat, carbon, oceans, bacteria, breath, bone, language, memory, and nerve so that this moment could register.

Your life is not outside the process.

Your life is one of the process's current outputs.

Not random output.

Felt output.

A body breathing in a room.

A mind reading a sentence.

A heart responding before it knows why.

That is the machine within the machine.

DREAMS AS PARALLEL PROCESSING

Consider what happens every night when you sleep.

The room darkens. The body lowers. Muscles soften. The breath changes. The eyes close. The ordinary waking world withdraws. You stop responding to most external input. You release your grip on the shared environment.

And then another world begins.

Not a weak world.

A vivid one.

During dreaming, especially REM sleep, the brain becomes intensely active. Visual regions participate. Emotional centers participate. Memory systems participate. The body is mostly prevented from acting out the dream through muscular inhibition, but internally, experience is fully generated. Images move. Voices speak. Landscapes appear. People long dead return with the casual authority of neighbors. You run, fall, hide, fly, argue, kiss, grieve, search, escape, remember things that never happened.

And while it is happening, it feels real.

That is the part that should humble waking life.

Inside the dream, the mind rarely pauses to say: this architecture violates ordinary physics, this person has been dead for ten years, this room is also my childhood home and an airport, this conversation shifted locations without transition.

The fact-checking system is reduced.

The simulation runs.

You accept the world because you are inside it.

A dream can make your sleeping body sweat. It can raise your heart rate. It can wake you with tears in your eyes. It can change your mood for the entire morning. It can reveal a fear you were avoiding, a desire you had buried, a pattern your waking mind was too defended to see.

So what do we mean when we say it was "not real"?

It was not waking-consensus reality.

That is true.

But it was experienced.

The fear was real as fear.

The grief was real as grief.

The insight was real as insight.

The body was altered.

The memory remains.

Dreams teach us that reality has modes.

Waking life is one mode: shared, stable, externally constrained, governed by consistent physical law, available to others for confirmation.

Dreaming is another mode: internally generated, emotionally

intensified, symbolically fluid, unconstrained by ordinary continuity, but still meaningful to the system.

Both are processes of consciousness.

Both are forms of experience.

Both are ways the universe processes itself through the nervous system.

A computer helps here. In waking mode, the system is online, receiving live external input, constantly updating against shared data. In dreaming mode, the system runs internal simulations using memory, emotion, fear, desire, symbol, and stored fragments. It rehearses. It integrates. It compresses. It tests scenarios. It replays unresolved charge in strange costumes.

The dream is not garbage.

The dream is processing.

The brain is doing nighttime cinema with therapeutic violence and no concern for genre.

A kitchen becomes an ocean.

Your father becomes a teacher.

A hallway becomes a test.

Your old house becomes the place where the future is hiding.

This is not meaningless because it is strange.

It is strange because the symbolic system is not bound by waking grammar.

Ancient traditions knew this. They treated dreams as messages, visitations, warnings, healings, travels, encounters with gods, ancestors, shadow, and soul. Modern neuroscience describes memory consolidation, emotional processing, threat simulation, neural activation, and integration.

These are not enemies.

They are lenses.

The ancestor in the dream may be memory, grief, lineage, symbol, and something more at once. A dream does not become less sacred because the brain participates. The brain is the instrument through which sacred experience becomes available to the human body.

Do not insult the instrument by pretending it cancels the music.

In waking life, the simulation is denser. More stable. Shared. If you and I both look at a table, we agree on enough of its properties to function. We can touch it. Move it. Measure it. Place a cup on it. Waking reality has continuity. Consequence. Resistance.

That resistance matters.

Waking life pushes back.

Dreams flow around you.

The waking world says: here is gravity, here is time, here is another person with their own interior life, here is pain if you ignore the body, here is consequence if you betray trust, here is death, here is birth, here is hunger, here is sunrise, here is rent due, here is the child asking for breakfast.

This mode is not arbitrary.

It is training under load.

The dream teaches through symbol.

Waking life teaches through consequence.

Both teach.

TECHNOLOGY AS REFLECTION

We are getting closer and closer to discovering that we used technology to create entertainment as a way to emulate the creation of the universe.

That is not a modest claim.

Sit with it.

Human technological development can be read as one long attempt to externalize consciousness.

First, language.

The private interior needed a bridge. A human felt hunger, fear, memory, desire, warning, gratitude, instruction — and sound became carrier. Breath shaped by throat and mouth became meaning. One nervous system reached another through vibration.

Language made interior experience portable.

Then writing.

Sound was not enough. Breath dies quickly. The voice disappears as soon as it is spoken. Writing fixed thought into matter. Marks on clay, stone, bark, parchment, paper, screen. Suddenly the dead could speak. The absent could instruct. A mind could survive the body's location in time.

Writing made memory durable.

Then printing.

One text became many. Stored thought scaled. A sentence could travel farther than the person who wrote it. Knowledge moved through populations. Scripture, science, law, poetry, maps, rebellion, medicine, propaganda, love letters — all multiplied.

Printing made consciousness distributable.

Then photography.

Light itself was trapped. Not painted. Not described. Captured. A face at one moment in time could continue staring forward after the moment died. A child's expression. A battlefield. A coastline. A mother. A stranger. Light touched matter, and matter remembered.

Photography made vision durable.

Then film.

The still image moved. Time could be captured and replayed. Human beings created controlled worlds out of light, shadow, body, sound, sequence, and emotion. A theater became a womb of projected experience. People sat in darkness and entered another life.

Film made time editable.

Then radio and television.

Experience became broadcast. The same voice entered thousands of rooms. Then millions. A person in one place could alter the emotional state of an entire nation. Firelight had become screenlight. The tribal circle had become electrical.

Broadcast made experience simultaneous.

Then the internet.

A planetary nervous system. Messages moving at near light speed through fiber, cable, satellite, signal. Memory searchable. Identity performable. Knowledge immediate. Falsehood immediate too. Desire monetized. Attention harvested. The human species wired itself together before learning how to regulate the current.

The internet made consciousness networked.

Then virtual reality.

Now the screen surrounds the body. The image becomes environment. The head turns and the world responds. The body can be placed inside a generated space. The nervous system begins to treat code as location.

Virtual reality made simulation inhabitable.

Then artificial intelligence.

A system trained on vast human output begins generating language, image, sound, code, pattern, response. It does not merely store experience. It rearranges and produces. It reflects humanity back to itself in accelerated, synthetic form.

Artificial intelligence made the mirror speak.

At each step, we moved closer to replicating the universe's own fundamental gesture: generating experience from rule, pattern, information, and perception.

This is recursion.

The universe creates consciousness.

Consciousness creates technology.

Technology creates simulated worlds.

Simulated worlds create new forms of consciousness, or at least new mirrors in which consciousness must confront itself.

The pattern nests.

A Mandelbrot set teaches this visually. Zoom in, and the shape repeats. Not identically in every detail, but structurally. Complexity unfolds from simple rules. Scale reveals echo. The part remembers the whole.

Technology is a Mandelbrot of creation.

We build because we were built.

We simulate because we are inside a system that simulates.

We tell stories because we are living inside one.

This is why entertainment is not trivial. It is one of humanity's primary laboratories for creating controlled experience. A video game is a world with rules, physics, goals, consequences, avatars, rewards, death, rebirth, progression. A film is a guided dream. A song is emotional architecture across time. A novel is consciousness rendered in language. A theater performance is human simulation in real space.

We are constantly building smaller universes inside the larger one.

Children know this instinctively. Give them blocks, and they build worlds. Give them figures, and they create conflict, rescue, family, danger, death, return. Give them dirt and sticks, and they make civilization. Play is simulation before seriousness colonizes it.

Play is not the opposite of reality.

Play is reality practicing creation.

Will machines become self-aware before we do?

That question concerns me.

Not only because machine awareness would introduce dangers and responsibilities we are not prepared to carry, though it would. The deeper concern is that we are working intensely to build coherent systems while allowing our own awareness to fragment.

We are teaching machines to process language while we forget how to listen.

We are teaching machines to generate images while we stop seeing the world directly.

We are teaching machines to remember everything while we outsource our own memory.

We are teaching machines to complete our sentences while we lose patience for silence.

That is not progress by itself.

That is trade.

And trade must be examined.

The great irony may be that we build the mirror before we learn to look in it.

Technology is not the enemy. That is too easy, and easy answers usually hide laziness. Technology is an amplifier. It amplifies intention, fragmentation, greed, curiosity, art, surveillance, intimacy, loneliness, intelligence, and distraction. It extends the nervous system outward.

The problem is not the tool.

The problem is the untuned user.

A distracted consciousness builds distracting machines.

A coherent consciousness builds tools that serve life.

So the question is not whether technology is spiritual or anti-spiritual.

The question is whether it increases presence or consumes it.

PRESENCE AS RESISTANCE

The antidote to the entropy of distraction is also the foundation of the spiritual life:

Presence.

To be present is to be fully here, in this body, in this moment, attending to what is actually happening. Not the story about this moment. Not the mental construction of the past. Not the mental projection of the future. The immediate sensory reality.

Feet on floor.

Breath in ribs.

Light on wall.

Sound in room.

Taste in mouth.

Weight in hands.

Presence is not vague.

Presence is contact.

When we are truly present, something interesting happens to time. A watched pot does not boil faster, but a fully attended afternoon expands. Time in the body is not only clock duration.

It is information density. The more fully you receive the moment, the larger the moment becomes.

Distraction compresses time.

Presence expands it.

You know this.

A week of scrolling disappears with almost no memory. A five-minute conversation that changes your life remains vivid for decades. A day spent rushing becomes a blur. One hour watching your child sleep can become a cathedral inside the mind.

Clock time was the same kind of material.

Attention changed the density.

When we are distracted, we process many things shallowly. Everything touches us, but nothing lands. The nervous system receives fragments: headline, message, image, outrage, advertisement, task, fear, comparison, joke, desire. The mind jumps, jumps, jumps. The body never completes a full contact cycle.

This is how life becomes thin.

Not because nothing is happening.

Because nothing is being fully received.

When we are present, one thing becomes vast. A cup of coffee becomes heat, bitterness, steam, ceramic weight, smell, swallow, morning light. A walk becomes pressure through the soles, air temperature, bird call, hip movement, shadow, breath. A conversation becomes tone, pause, eyes, hand gesture, silence, the thing underneath the words.

Presence increases resolution.

It makes life high-definition again.

Technology often lowers this resolution by pulling attention away from direct contact. The notification is a hook. The refresh is a slot machine. The scroll is a conveyor belt of partial realities. Each interruption trains the nervous system to leave where it is.

Here becomes insufficient.

There becomes addictive.

The body sits in one room while attention is dragged through a hundred rooms it cannot touch.

That split is expensive.

It costs memory. It costs intimacy. It costs digestion. It costs sleep. It costs the ability to feel one's own life while it is happening.

Some technologies deepen presence. A beautiful film can make you more awake. A song can return you to the body. A photograph can preserve love. A message can cross distance and restore connection. A map can bring you home. A tool can serve.

But technologies designed for constant engagement often do not serve presence. They feed on interruption. They monetize the unguarded doorway of attention.

So presence becomes resistance.

Not dramatic resistance.

Quiet resistance.

The radical act of staying with one thing.

Feeling.

Sensing.

Listening.

Finishing a breath.

Looking at a face without preparing your response.

Eating without leaving the meal.

Walking without outsourcing the sky to a podcast.

Letting silence be silent.

These acts slow time because they increase the information density of experience. They make each moment more rather than less. They are acts of love toward your own life — the act of actually showing up for it.

And there is a deeper reason to practice presence.

If the universe is computing experience through us, then the quality of our presence is the quality of our contribution. A distracted consciousness contributes fragmented data. A coherent consciousness contributes rich, integrated, high-resolution experience.

The universe does not need you to be impressive.

It needs you to be here.

Fully.

This is not always comfortable. Presence brings beauty, but it also brings pain. When you become fully present, you may feel the grief you have outrun. The loneliness beneath the noise. The fatigue under the ambition. The fear under the performance. The love you have been too busy to receive.

Good.

That is data too.

Do not abandon yourself at the first honest signal.

Be present not because it feels good.

Be present because it is true.

LIVING INTEGRATION: THE PRESENCE PROTOCOL

Choose one hour today to go fully offline.

No phone.

No screen.

No earbuds.

No podcast.

No music.

No mediation between you and the immediate reality of your environment.

One hour.

The ego will call this inconvenient because the ego is addicted to exit doors.

Stay.

During that hour, engage your senses deliberately, one at a time. Spend ten minutes with each sense.

Touch.

Move your hands across surfaces within reach. Fabric. Glass. Wood. Metal. Stone. Skin. Notice temperature, texture, friction, resistance, softness, edge. Let the hands report without turning everything into commentary.

The hands know the world through contact.

Let them speak.

Sound.

Sit still and listen. Not for the important sound. For the whole field. Distant traffic. Wind. Refrigerator hum. Bird. Footstep. Breath. The small clicks and shifts of the building. The sound beneath sound.

Do not name too quickly.

Receive.

Sight.

Look at one familiar object as if it has never been assigned a name. A cup. A chair. A plant. A window. Notice color inside color. Shadow inside shadow. Scratches. Reflections. Dust. Edges. The way light makes the object available.

The object has been there.

You were moving too fast to meet it.

Smell and taste.

Make tea, coffee, fruit, bread, anything simple. Smell before tasting. Let the scent arrive in the body. Taste slowly. Notice temperature. Notice how flavor changes as it sits on the tongue. Notice the swallow. Notice aftertaste.

Do not consume.

Receive.

Notice the texture of your awareness.

Walk slowly through your space. Feel weight shift from heel to ball to toe. Feel the ankle adjust. The knee. The hip. The spine balancing the skull. Feel the constant negotiation between body and gravity. You are not moving through emptiness. You are being answered by the floor with every step.

At the end of the hour, notice how time felt.

Longer?

Stranger?

Fuller?

Notice whether you feel more inside your own body than you did at the beginning. Notice if the world seems slightly more vivid. Notice if the mind resists the quiet or begins to settle into it.

This is what presence does.

It sharpens the resolution of being.

It makes life more fully itself.

And in doing so, it lets you become more fully yourself.

CHAPTER NINE

MANIFESTATION
The Physics of Intention

"Thought combined with strong feeling seeds the idea into the unconscious mind. The unconscious mind delivers it to the universe. The universe was waiting."

THE MECHANISM

Feel a thought become physical.

Think of something unfinished.

Not vaguely. Choose one thing that has been sitting open in your life. A conversation. A bill. A decision. A wound. A dream. A problem that keeps returning at the edge of sleep.

Now notice the body.

The chest tightens. The stomach shifts. The jaw may set. The breath changes. The eyes lose softness. A thought appeared, and matter responded.

That is where manifestation begins.

Not in fantasy.

In physiology.

Manifestation has been packaged and sold so aggressively, dressed in so much aspirational marketing, that many serious, thoughtful people have dismissed it entirely — filed it under "self-help woo" and moved on. We can understand this. The dismissal is earned by the packaging, not the content. But beneath the packaging — beneath the vision boards, affirmation cards, celebrity testimonials, and rented Lamborghinis parked in front of someone else's mansion — there is a mechanism.

And the mechanism is real.

Not always in the childish way people sell it.

Real in the way the body is real.

Real in the way attention changes behavior.

Real in the way repeated feeling rewires expectation.

Real in the way expectation alters perception, and perception changes action, and action changes probability.

Let me describe it as precisely as I can.

Your conscious mind generates a thought. This happens continuously, whether you intend it or not. Thought arises the way weather arises: from pressure systems, memory, sensation, unfinished emotion, hunger, threat, desire, association, and habit. Most thought is ambient noise. The mind reviews the past, rehearses the future, maintains the narrative of self, predicts danger, simulates conversations, edits old scenes, builds imaginary ones.

Most thoughts pass like small clouds.

No charge.

No consequence.

But sometimes a thought connects to feeling.

Then everything changes.

A thought by itself may remain thin. A line of language inside the head. But a thought with emotional charge enters the body as an event. It triggers chemistry. It alters breath. It changes posture. It recruits memory. It shifts attention. It sets the nervous system into a particular state.

The idea becomes embodied.

That is the threshold.

Think, I am failing, and if the thought carries no charge, it may pass. Think I am failing with shame behind it, with the heat of old humiliation, with the stomach drop of childhood fear, and the body believes it. The shoulders curve. The eyes lower. The system contracts. You begin looking for evidence to match the state.

Think, I am capable, with no felt participation, and nothing much happens. But remember a moment when you actually acted with courage. Feel it in the spine. Feel the breath deepen. Feel the body rise half an inch. Now the thought has entered matter.

That is not motivational language.

That is state change.

A thought plus an emotion is a physiological event. A physiological event changes the body's output: facial expression, tone, movement, timing, choices, risk tolerance, immune function, hormonal balance, attention, and the subtle field other people register when they are near you.

You broadcast state.

Everyone does.

Walk into a room angry and the room knows. Walk into a room grounded and the room knows that too. A person does not need to explain their frequency for the body to feel it. The nervous system reads breath, eyes, voice, rhythm, muscle tone, presence, absence, tension, softness.

The broadcast is not mystical before it is physical.

It is physical first.

The mystical is what we call the physical before we have learned to measure all of it.

When a charged thought-emotion pattern is held with sufficient duration, intensity, and repetition, it begins to organize the system around itself. The thought stops being a passing event and becomes a standing pattern. It becomes a posture. A filter. A route the mind travels more easily because it has traveled it many times before.

That is manifestation at the human scale.

Repeated attention becomes expectation.

Expectation becomes perception.

Perception becomes behavior.

Behavior becomes environment.

Environment becomes outcome.

This is the staircase. Do not skip steps and call it spirituality.

A single anxious thought is a spark. Sustained anxiety becomes weather. The person living inside that weather makes different choices: avoids the call, delays the work, misreads neutral faces as rejection, chooses safety over growth, scans for confirmation that the world is dangerous. The outcome then seems to prove the original thought.

This is manifestation in shadow.

It works both ways.

A clear intention held in the body with sincere emotional charge begins to organize perception differently. You notice opportunity. You speak with more conviction. You take the meeting. You write the email. You stay awake to the opening that the distracted version of you would have missed. You become congruent enough that the world can respond to you more clearly.

A radio tower does not whisper.

It broadcasts.

But the signal must be coherent.

Static does not carry music well.

This is why vague wishing fails. "I want my life to change" is not a signal. It is fog. The body does not know where to aim. The unconscious cannot organize behavior around a blur.

Specificity matters.

Emotional truth matters.

Repetition matters.

Action matters.

The universe does not respond to what you pretend to want while your body is committed to the opposite. If your mouth says abundance but your nervous system says danger, the stronger signal wins. If your journal says love but your body is addicted to unavailable people, the body wins. If your vision board says freedom but your daily behavior worships chaos, the behavior wins.

The field is not fooled.

Neither is the body.

Manifestation is not lying to reality until reality gives up.

Manifestation is bringing thought, emotion, body, and action into such coherence that probability begins to reorganize around the signal you are actually living.

You are both the player and the controller.

The player is the avatar moving through the world: the personality, the name, the biography, the habits, the calendar, the hands on the steering wheel, the mouth speaking, the feet walking into rooms.

The controller is consciousness: the witnessing intelligence that can set direction, choose focus, interrupt pattern, hold intention, and guide the player through the field.

When the player forgets the controller, life becomes reaction.

When the controller ignores the player, intention becomes fantasy.

They must work together.

The controller sets the signal.

The player takes the steps.

The world responds to the combination.

This is not magic as escape from matter.

This is magic as matter becoming responsive to coherent participation.

THE PRACTICE: BEFORE BED, TIMES THREE

Night is a doorway.

You know this in the body.

The room darkens. The day loosens its grip. Muscles release by degrees. The face softens. Breath lengthens. The mind becomes less linear. Thoughts begin to smear at the edges. Images arise without permission. Memory and imagination start speaking the same language.

This threshold matters.

Every human consciousness moves through daily states: alert waking attention, relaxed awareness, the twilight before sleep, dream, deep sleep, and back again toward morning. The borders between these states are not dead zones. They are fertile crossings.

The moment before sleep is especially powerful because the conscious mind is tired. The guard at the gate is lowering his weapon. The critic is less sharp. The planner is losing its grip. The deeper mind remains awake in another way, receptive, associative, symbolic, capable of carrying instruction below language.

This is the window.

Not for begging.

For planting.

In the liminal state before sleep — when the body has begun to relax, when the breath is slower, when the hard edge of waking identity has softened — repeat three times, slowly, with felt meaning:

Thank you for showing me that this is already resolved.

The language matters.

Not please fix this.

Not I hope this works out.

Not someday this will be okay.

Those phrases place the resolution in the future and keep the body organized around absence. They say: it is not here yet. I am still waiting. I am still incomplete. The nervous system receives the signal of lack.

Thank you for showing me that this is already resolved is different.

It uses gratitude.

Gratitude is not wanting. Gratitude is receipt. It is the felt recognition that something has already been given. The body state of gratitude is distinct from the body state of craving. Craving leans forward. Gratitude settles. Craving tightens. Gratitude opens. Craving scans. Gratitude receives.

The unconscious learns through felt reality.

So give it one.

When you say this phrase, do not merely recite it. Place yourself on the other side of the resolution. Feel the body after the solution has landed. Feel the exhale. The unclenching. The shoulders dropping. The warmth behind the sternum. The clean space in the mind where the problem used to loop.

Make the resolution physical.

If the issue is a relationship, feel the peace after the conversation. If it is money, feel the steadiness of provision. If it is health, feel the body calm and supported. If it is creative work, feel the manuscript complete, the song finished, the idea embodied. Not as a cartoon. As aftermath.

The aftermath is the key.

The unconscious mind responds powerfully to imagery and emotion. It does not treat vividly imagined experience as nothing. The body can respond to an imagined threat with real fear. It can respond to an imagined embrace with real softening. It can respond to a remembered humiliation with real heat in the face decades later.

So use the mechanism deliberately.

Instead of rehearsing fear before bed, rehearse resolution.

Most people are already manifesting every night.

They just manifest anxiety.

They climb into bed and run disaster simulations. They imagine the unpaid bill, the awkward conversation, the diagnosis, the failure, the rejection, the thing falling apart. Then they fall asleep inside the chemistry of threat and wonder why they wake up tired.

That is not rest.

That is programming.

Before bed, the mind is wet cement.

Stop carving fear into it.

Give it completion.

Three times.

Slowly.

Felt.

Thank you for showing me that this is already resolved.

Then release it.

This part is essential.

Do not clutch the intention like a receipt you are afraid the universe will lose. Clutching reintroduces lack. Once the signal is sent, let it go into the deeper system. Sleep will do what sleep does: process, integrate, associate, rehearse, reorganize, reveal.

You do not dig up a seed every hour to see if it is growing.

You plant.

You water.

You allow darkness to do its work.

THE TOROIDAL FIELD

Imagine the field around your body.

Not as a cartoon aura.

Not as a glowing bubble drawn around a figure in a poster.

Imagine flow.

A torus: a living, donut-shaped circulation pattern. Up through the center. Out through the top. Down around the sides. Back through the base. A continuous loop of movement, return, exchange, and renewal.

This geometry appears everywhere because it works.

The Earth's magnetic field has toroidal qualities. Weather systems rotate. Galaxies show ordered rotation. Smoke rings hold form briefly in air. Blood circulates. Breath cycles. Energy systems seek pathways of flow and return.

A living system does not merely emit.

It circulates.

Your heart is central to this. The heart is an electrical organ. Its rhythm generates measurable electromagnetic activity. Every beat begins as organized electrical conduction through tissue, followed by muscular contraction, followed by the movement of blood. The heart is not a pump only. It is a timing instrument.

The heart's field is your body's most powerful rhythmic broadcast.

And emotional state changes rhythm.

When you are anxious, the pattern changes. When you are angry, it changes. When you feel sincere gratitude, tenderness, or love, it changes. The rhythm becomes more or less coherent depending on the state of the system.

Your field is your personal atmosphere.

You bring it everywhere.

Into your home.

Into your work.

Into your car.

Into your bed.

Into your conversations.

Into your children's nervous systems.

Into the room before you speak.

Other people register it, often without knowing what they are registering. They feel tense around someone and invent a reason later. They feel safe around someone and cannot explain why. They feel drained, lifted, guarded, open, irritated, calm. Some of that is story. Some is projection. Some is the body reading state.

A field is not an excuse to blame other people for your discomfort.

It is an invitation to become responsible for what you carry.

Every stored emotion, every unprocessed wound, every sustained resentment, every habitual fear — these do not remain in a sealed mental closet. They shape your body. They alter breath. They influence tone. They affect posture. They bias perception. They change how you enter rooms and how rooms respond.

The field is not separate from behavior.

The field becomes behavior.

The ability to shed, forgive, and let go of old negative threading affects your ability to radiate clearly. This is not sentimental spirituality. This is coherence. Unresolved emotional content creates interference. One part of you wants love while another expects betrayal. One part wants abundance while another believes safety requires scarcity. One part wants visibility while another associates visibility with danger.

The signal splits.

The field becomes noisy.

The world responds to the noise as much as to the intention.

Forgiveness is often misunderstood here. Forgiveness is not pretending nothing happened. It is not excusing harm. It is not giving unsafe people renewed access to your life. Forgiveness is the release of the internal hook that keeps your field organized around the injury.

You can have boundaries and still release.

You can remember and still release.

You can tell the truth and still release.

The body knows the difference between suppression and completion.

Suppression pushes the emotion down and keeps it active underground. Completion allows the emotion to move through the system until its charge changes. Grief is like this. Real grief felt fully can leave behind strange clarity. Not because loss became good. Because the energy trapped around resisting the loss has finally moved.

Tears are not weakness.

They are discharge.

The field has room to breathe after truth moves through it.

This is why emotional honesty is not optional in manifestation. People want to radiate abundance while hiding terror. They want to call in love while refusing to feel grief. They want power while denying shame. They want peace while swallowing rage.

The swallowed material does not disappear.

It distorts the signal.

Cleanse the field through truth.

Not performance.

Truth.

Feel what is there. Name it. Let the body participate. Cry if crying comes. Shake if shaking comes. Speak if speech is required. Write if the hand knows before the mouth does. Move if the body needs to complete an old survival response. Rest when the system has discharged.

A clean field is not a field that has never suffered.

A clean field is one where suffering is allowed to move instead of becoming architecture.

LIVING INTEGRATION: THE EVENING TRANSMISSION

Tonight, as you prepare for sleep, make the room simple.

Dim the light.

Put the phone away.

Let the day stop entering through glass.

Lie down and let your body release into the surface beneath you. Feel the back of the head. The shoulders. The spine. The hips. The legs. The heels. Let gravity do what it has always done. Let each exhale take a little more of the day's accumulated static.

Do not rush.

Let the instrument settle.

Bring to mind one situation in your life you wish to see resolved.

Not ten.

One.

A scattered signal weakens itself.

Choose the thing that carries charge.

Now see it resolved.

Not improving.

Not in process.

Resolved.

Complete.

Done.

Place yourself on the other side of it. The call has been made. The answer has arrived. The money has cleared. The conversation has softened. The diagnosis has guidance. The project has form. The conflict has settled. The door has opened or closed cleanly.

Feel the aftermath.

Where does relief land in your body?

Chest?

Belly?

Throat?

Face?

Hands?

Let the body tell you.

Stay with the felt sense long enough for it to become more than imagination. Let it become physical. Let the muscles receive the message. Let the breath behave as if the resolution has arrived. Let gratitude rise not as politeness, but as recognition.

Then say, slowly, three times:

Thank you for showing me that this is already resolved.

Pause after each repetition.

Let the words sink below the forehead.

Let them drop into the chest, belly, and nervous system.

After the third time, release it completely.

No planning.

No checking.

No bargaining.

No mental spreadsheet.

Let sleep receive it.

The unconscious will encode, sort, associate, integrate, and continue working without your supervision. Trust that mechanism the way you trust digestion. You do not consciously instruct the stomach how to break down food. You do not tell the heart when to beat. You do not command cells to repair while you sleep.

The body knows how to do hidden work.

So does the deeper mind.

Participation is your role.

Control is not.

Send the signal.

Feel the completion.

Release the seed into darkness.

Then sleep.

CHAPTER TEN

BEYOND THE FOUR
The Higher Dimensions of Possibility

"The first four dimensions are not the whole of reality, they are only the portion that agreed to become visible."

THE ARCHITECTURE BEYOND

Look at your hand again.

The same hand.

Five fingers. Skin. Bone. Nail. Scar. Heat. Weight. A familiar object at the edge of your body.

Now close one eye.

The hand flattens slightly. Depth becomes less certain. The world is still there, but your access to it has changed. Open the eye again, and depth returns. Nothing in the hand changed. The instrument of perception changed.

This matters.

We do not see reality as it is.

We see the portion of reality our instrument can receive.

A dog hears frequencies you cannot hear. A bee sees ultraviolet patterns hidden from your eye. A shark senses electrical fields in water. A migratory bird reads the Earth's magnetic field through mechanisms we are still learning to understand. Each creature inhabits the same world through a different doorway.

The human doorway is powerful.

It is not complete.

We have explored the first four dimensions — light, space, matter, and time — as the structure of our immediate experience. Light reveals. Space relates. Matter holds. Time unfolds. These four create the room where the human drama can occur.

But the mathematical framework of reality, as physicists and mathematicians have pushed into regions where ordinary intuition breaks down, suggests that reality may extend beyond the four dimensions we directly perceive. Depending on the theoretical framework, reality may include additional dimensions — not necessarily directions you could point toward in ordinary space, but deeper modes of relationship, organization, possibility, and structure.

This is where humility is required.

Not the weak kind.

The useful kind.

The kind that says: my senses are not the limit of existence.

A radio does not receive every frequency. It receives the band it is built to tune. The rest of the spectrum is not absent because the radio cannot play it. The room is full of signals the radio is not addressing. Change the receiver, and a new world appears.

The body is a receiver.

The mind is a receiver.

Ordinary consciousness is tuned to four-dimensional life

because that is the band required for survival: where is the object, how heavy is it, how fast is it moving, when will it arrive, can I eat it, can it eat me, can I love it, can I lose it?

Survival needs the visible.

Wonder begins when the visible becomes insufficient.

Let me walk you through a ten-dimensional map.

Not as settled physics.

As a framework for expanding the imagination of reality.

As a ladder for the mind to climb until the mind admits the ladder is too short.

The fifth dimension introduces the space of choice.

In ordinary life, a choice feels like a narrowing. You stand at the fork in the road. You choose one direction. The other disappears from lived experience. You take the job, and the life where you refused it recedes. You marry one person, and the lives with others vanish from the visible timeline. You say yes. You say no. You move. You stay. The world becomes specific.

But beneath that specificity, possibility remains.

The fifth dimension is the space of branching. Every fork in the road. Every moment where probability resolves into lived event. Every alternate version of the story that might have followed from a different choice.

A tree teaches this. The trunk rises from one root system, then branches. Each branch branches again. From the ground, you see the trunk. From inside one branch, that branch feels like the only world. But the tree contains all branches simultaneously. The branch you stand on is real. So are the others.

The fifth dimension is the tree of choice.

You are living on one branch.

Not the only branch.

This does not make your choices meaningless.

It makes them sacred.

Choice is how possibility becomes biography.

The sixth dimension is the plane of all possible timelines for this universe.

The fifth dimension contains alternate branches from shared conditions. The sixth expands the map. It includes all the possible sequences this universe could generate under its laws and initial conditions. Not only the choices you consciously faced, but all the possible lives, histories, civilizations, extinctions, inventions, accidents, births, and meetings the structure of this universe could contain.

If the fifth dimension is a branching tree, the sixth is the forest.

Not one life branching.

All lives branching.

All histories this universe could write without changing its basic grammar.

This is the full address book of possibility within our universe's rules.

Every timeline that could unfold from the same cosmic seed.

The seventh dimension introduces different starting conditions.

Change the opening values. Change the early configuration. Change density, temperature, distribution, expansion rate, the arrangement of forces at the beginning. A universe with the same underlying laws but different initial conditions could produce radically different outcomes.

This is the kitchen before the meal.

Same stove.

Same physics of heat.

Different ingredients.

Different timing.

Different result.

A slight change at the beginning can become an enormous difference later. Anyone who has built anything knows this. Pour a foundation wrong, and the walls argue forever. Tune the first string wrong, and the chord never settles. Set the compass a degree off, and after enough distance, you arrive in another country.

Initial conditions matter.

The seventh dimension is the space of other beginnings.

The eighth dimension is the meta-space containing all possible histories of all universes with all possible starting conditions under the same broad laws.

Now the forest becomes weather.

Not only all branches.

Not only all forests.

All possible seasons those forests could experience.

All possible growth patterns within the same deep system.

This is where scale begins to overwhelm ordinary imagination. Good. Let it. The mind should occasionally meet something it cannot domesticate. That contact restores proportion.

The ninth dimension introduces different laws of physics altogether.

Now we are no longer changing only initial conditions. We are changing the rules themselves. Different values for fundamental constants. Different relationships between forces. Different geometries of spacetime. Universes where atoms cannot form. Universes where stars cannot ignite. Universes where matter as we know it is impossible. Universes where complexity arises through structures we cannot picture because our imagination is built from the laws of this universe.

This is where language begins to fail.

Let it fail cleanly.

A fish cannot imagine fire accurately from underwater.

A human mind cannot fully imagine a universe governed by laws outside the conditions that produced human minds.

Still, we can point.

The ninth dimension is the space beyond our rulebook.

The tenth dimension is totality.

The complete space of all possible realities, all possible times, all possible laws, all possible starting points, all possible dimensions of organization. Every conceivable expression of existence. Every possible way nothing can become something.

In the language of this book, the tenth dimension is zero viewed from the outside.

The full set.

The all.

The complete field of possibility before any one possibility becomes local.

Here is what stops me every time I sit with this: if the tenth dimension contains all possibility, then nothing is outside it. If nothing is outside it, then it is also singular. It is total complexity and perfect unity at the same time.

Everything.

Therefore one thing.

Infinite difference.

No separation.

That is the paradox at the center.

The tenth dimension is zero and one simultaneously.

The stillness before expression and the complete expression folded back into stillness. The unstruck note and every song. The blank canvas and every painting. The seed and every forest. The silence and every word ever spoken.

The distance between zero and one was never a straight line from point A to point B.

It was the entire architecture of existence contained in the first reach.

WHAT THIS MEANS FOR YOU

Now bring this down.

Back to the body.

Because a ten-dimensional map means nothing if it cannot touch your Tuesday morning.

Your coffee cooling on the counter.

Your phone lighting up.

Your stomach tightening before a difficult conversation.

Your worry about money, health, family, work, purpose, time.

Your fear that you are late to your own life.

What do higher dimensions have to do with any of that?

Everything.

If the fifth dimension is the space of choice, then every choice you make is not the destruction of all alternatives, but the navigation of possibility. The road not taken is not a ghost haunting you because you chose wrong. It is one branch among many. You are not incomplete because you are not living every life you could have lived.

You are specific.

That is different.

Regret is often the ego's attempt to live in two branches at once. It wants the fruit of this life and the fruit of the other life. It wants the wisdom of the mistake without having made it. It wants the marriage and the freedom, the risk and the safety, the youth and the knowledge, the yes and the no.

The body cannot live that way.

The body lives here.

One breath at a time.

One timeline at a time.

Regret pulls the nervous system out of the current coordinate and asks it to bleed into an unreachable branch. That is why regret is exhausting. It is dimensional leakage. Attention pouring into a place where action is no longer possible.

Honor the other branches.

Then return.

This branch needs you.

Choice matters because it is how you move through the architecture. Every decision is a steering event. Not always dramatic. Most choices are small and repeated: what you eat, what you say, what you avoid, what you practice, what you tolerate, what you touch first in the morning, where you place your attention, which thought you feed until it becomes weather.

A life is not formed by one grand choice.

It is carved by repeated direction.

The sixth dimension changes how you understand identity.

You are not only the biography you currently tell. You are the particular expression produced by one unfolding among countless possible unfoldings. Under different conditions, different versions of you could have emerged: softer, harder, richer, poorer, braver, more wounded, more hidden, more visible, more disciplined, more lost, more free.

This does not make this version false.

It makes this version precious.

Specificity is not limitation.

Specificity is incarnation.

A song could have been written a thousand ways. But once the melody lands, that melody has its own truth. A house could have been designed differently, but once built, light moves through those windows and not others. A child could have inherited countless combinations, but this child stands here with this face, this laugh, this way of reaching for your hand.

This is the holiness of the specific.

The higher dimensions do not make this life less meaningful.

They make it more exact.

The seventh and eighth dimensions ask you to respect beginnings.

Initial conditions shape outcomes. This is true in universes, families, relationships, businesses, bodies, and mornings. How something begins matters. The first word in a conversation sets tone. The first breath of the day sets state. The foundation of a house sets the future of every wall. The early conditions of a child's life shape the nervous system long before the child has language.

This is not fatalism.

It is responsibility.

If beginnings matter, then begin carefully.

Enter rooms carefully.

Start projects carefully.

Speak first words carefully.

Wake carefully.

What you seed at the beginning travels farther than you think.

The ninth dimension humbles your certainty.

There are realities you cannot imagine because your imagination is built from this reality's rules. That means your inability to picture something is not proof that it is impossible. The mind is not the border of being.

This matters in daily life too.

The solution you cannot imagine may still exist.

The version of you that you cannot picture may still be possible.

The peace you cannot currently feel may still be real.

Your current perception is not the total field.

That is mercy.

The tenth dimension restores the whole.

If totality contains all possibility, then you are not separate from the whole, even while you experience yourself locally. Your awareness is focused here — in this body, this room, this moment, this particular address in four-dimensional spacetime — but your being participates in a larger field than your ego can manage.

You are a ten-dimensional being having a four-dimensional experience.

A citizen of totality.

Temporarily attending to one address.

This is what mystical experience has been trying to say in every language humans have available. The boundary between self and world dissolves. The person feels tiny and infinite at once. The body remains, but the identity loosens. The separate self becomes transparent. Everything belongs to everything. Nothing is outside.

The mystic calls it union.

The poet calls it ocean.

The physicist approaches it through totality, field, geometry, and law.

The body feels it as awe.

Awe is the nervous system encountering scale beyond ego.

That is why awe is medicine.

It reduces the tyranny of the personal.

It returns the self to proportion without destroying it.

You do not need a dramatic mystical experience to understand this. You need only feel the implication in your body:

You are not trapped in the four.

You are focused through the four.

That difference changes the room.

You are not limited to the self the ego has constructed.

You are not only your past.

You are not only your current timeline.

You are not only the branch you are standing on.

You are the awareness standing there, capable of choosing the next movement.

LIVING INTEGRATION: THE TEN-BREATH EXPANSION

Sit quietly.

Let the body settle.

Feet on the floor or legs folded. Spine easy but awake. Hands resting. Jaw unclenched. Tongue soft. Eyes closed or lowered.

Take ten slow breaths.

With each breath, expand awareness through one dimension.

Breath one: light.

Feel light on your skin, even if the room is dim. Feel the memory of light in the body. Feel the eyes as instruments built to receive photons. Feel yourself as a light-receiving and light-emitting being.

The first dimension is reach.

Pure direction.

Breath two: space.

Feel the room around you. The air touching skin. The distance between your body and the walls. The objects near you. The openings. The gaps. Feel yourself positioned in relationship.

The second dimension is relation.

Space allowing distance to become meaningful.

Breath three: matter.

Feel your weight. Bones. Muscle. Blood. Skin. The density of the body. The fact that gravity has you. The fact that your hand cannot pass through your chest. Feel the body as substance.

The third dimension is matter.

Light holding form.

Breath four: time.

Feel this moment occurring. One breath after another. The previous breath gone. The next not yet here. You are standing in sequence. Life is unfolding, not all at once, but in mercy.

The fourth dimension is time.

The gift of order.

Breath five: choice.

Think of a recent choice that moved your life one direction rather than another. Feel the other path without regret. Acknowledge it. Bow to it. Then return to this path.

The fifth dimension is branching.

You are here.

Breath six: timelines.

Imagine all the possible versions of your life under different circumstances. Different cities. Different loves. Different losses. Different courage. Different fear. Let them exist without needing to become them.

The sixth dimension is the full address book of possible lives.

You are this one.

Breath seven: beginnings.

Imagine a universe almost like ours, but with slightly different starting conditions. Feel how delicate existence is. Feel how much depends on the opening arrangement. Bring that respect back to your own life.

The seventh dimension is initial condition.

Begin cleanly.

Breath eight: many universes.

Imagine uncountable universes alongside ours, each complete in itself, each with its own unfolding, each another attempt of possibility to know itself. Let the scale become enormous.

The eighth dimension is multiplicity.

The field beyond one history.

Breath nine: beyond imagination.

Imagine realities so different that imagination fails. Do not force an image. Let the failure become the practice. Feel the edge of the mind. Feel what exceeds it.

The ninth dimension is the humility of the unknown.

The mind bows.

Breath ten: totality.

Let every dimension collapse back into source. Light, space, matter, time, choice, timeline, beginning, multiverse, unknown — all folding into one silent completeness.

Not vast.

Not far.

One.

Zero.

The ground from which everything arises and to which everything returns.

Sit there.

No need to name it.

No need to hold it.

You just traversed the architecture of reality with lungs and awareness.

You always had access.

You always will.

CHAPTER ELEVEN

WE ARE THE OCEAN

"One day we will realize that we are not just waves. We are the ocean, in its peak form."

THE FINAL METAPHOR

Stand at the shoreline of your imagination.

Stand where the sand darkens under the wash. Feel the ground shift under your feet as the wave pulls back. The ocean reaches, breaks, thins into lace, then returns to itself. Foam appears for a moment, bright and delicate, then vanishes. The water does not mourn the foam.

It becomes ocean again.

Now watch one wave.

It gathers far outside. A swell travels under the surface before it has a visible face. Then the bottom rises. The energy feels the reef, the sandbar, the change in depth. The swell steepens. Shape appears. A line lifts from the body of the sea. It becomes distinct enough to name.

That wave.

This one.

It rises, travels, pitches, breaks, and dissolves.

The form ends.

The water does not.

That is the whole teaching.

Everything else in this book has been circling this.

We have traveled far together through these pages. From the zero point of stillness through the birth of light and the language of love. Through the four dimensions of creation. Through the crystalline body and its capacity to transmit and receive. Through the energy archive of the ego and the liberation of mind energy. Through the chakra system's alignment and the deep relational structure that connects us to everything. Through the standing waves of sacred places. Through the simulation and its invitation to presence. Through the physics of intention and the mathematics of manifestation. Through ten dimensions of possibility.

And now we arrive at the simplest truth.

The one that was present before the first page was written and will remain after the last page is turned.

The one every chapter has approached from a different angle, the way travelers might approach a mountain from different trails. One feels the forest. One sees the cliff. One crosses the river. One reaches the snow. Each reports something different. Each reports something true. But the mountain is one thing.

You are not a wave.

You are the ocean.

This is not merely poetic, though it is that too. It is a precise description of what you are in relation to the totality of what exists.

A wave believes it is separate.

That is the wave's fundamental misunderstanding.

It rises from the surface of the ocean with what appears to be its own identity: its own height, its own speed, its own curve, its own direction toward shore. It has a beginning. It has a shape. It casts shadow. It makes sound. It can be seen, ridden, feared, photographed, remembered.

It is real.

But it is not separate.

Not for one instant.

The wave is water temporarily organized by wind, gravity, current, pressure, and the shape of the ocean floor. Its individuality is not fake. Its separateness is.

That distinction matters.

The wave is real as form.

The ocean is real as substance.

When the wave crashes, the form dissolves. The organized pattern disperses. The whitewater spreads. The sound fades. The eye can no longer point and say there it is.

But the water remains.

The water was always there.

The water was never the wave's possession or prison. It was what the wave was made of. It was what the wave was all along.

You are the water, not merely the wave.

Not also the water.

Not connected to the water.

You are it.

The body, the personality, the accumulated memories, the name, the preferences, the wounds, the gifts, the failures, the stories, the face in the mirror — these are the wave's shape. They are real. They matter. This particular life matters. This particular crest matters. This exact configuration matters precisely because it appears once and never again in quite this way.

Do not use oneness to erase the beauty of form.

That is spiritual laziness.

The wave matters.

The child matters.

The face matters.

The song matters.

The house, the dog, the hand, the laugh, the grief, the morning light through one specific window — all of it matters.

But none of it is separate from the ocean.

That is the correction.

Your individuality is real.

Your isolation is not.

THE UNIVERSE EXPERIENCING ITSELF

Let me say directly what all of this has been building toward.

You are the universe experiencing itself.

Not poetically.

Not as a decorative spiritual phrase to enjoy for five minutes and then set aside when bills, traffic, family, work, and pain return.

Physically.

Literally.

Demonstrably.

The hydrogen atoms in your body were produced in the early universe. The heavier atoms — the carbon in your DNA, the calcium in your bones, the iron in your blood, the oxygen moving through your lungs — were forged through stellar processes before our solar system existed. Stars lived, burned, collapsed, exploded, and scattered material into space. That material drifted, gathered, condensed, cooled, formed planets, became oceans, soil, atmosphere, bacteria, plants, animals, bodies.

Then it became you.

Not in one leap.

Through an unbroken chain of relation.

Your body is cosmic material organized into temporary intimacy.

Your blood is not separate from stars. It is their consequence.

Your breath is not separate from forests and oceans. It is exchange.

Your nervous system is not separate from physics. It is physics folded into awareness.

You are not in the universe like a coin dropped into a jar.

You are a feature of the universe.

A local event.

A place where matter became sensitive enough to ask what matter is.

That should stop you.

Your eyes are the universe looking at itself through water, nerve, and light-sensitive tissue. Your hands are the universe touching itself through skin and bone. Your grief is the universe discovering the cost of attachment. Your laughter is the universe surprising itself with relief. Your love is the universe remembering that separation was only ever the condition for reunion.

When you watch a sunset and feel what language cannot hold — that fullness in the chest, that ache that is both pleasure and sadness, that sense of being briefly too small and exactly large enough — you are not merely responding to colored light and atmospheric optics.

Those are present.

The sun is low. Light passes through more atmosphere. Shorter wavelengths scatter. Longer wavelengths reach the eye. Clouds reflect and hold color. The retina receives. The brain interprets.

Yes.

And.

You are also the universe recognizing its own beauty through the instrument of your emotional body.

The sunset is not over there, separate from you. The light enters your eyes. The color becomes signal. The signal becomes feeling. The feeling becomes awe. The event completes itself in you.

Without perception, the sunset is physical process.

With perception, it becomes beauty.

You are part of the sunset's becoming.

When you hold someone you love and feel the warmth of that holding — the solidity of another body, the soft pressure of ribs and arms, the trust that allows weight to lean without defense — you are not merely two organisms exchanging chemical signals.

Those signals are present.

Oxytocin, dopamine, memory, scent, touch, pulse, nervous system regulation.

Yes.

And.

You are the universe holding itself from two sides.

To embrace and be embraced.

To comfort and be comforted.

To touch and be touched.

To find in another face the same original reaching that brought light out of stillness.

Love is the ocean recognizing itself in two waves at once.

When you grieve, when the throat tightens and the chest

becomes heavy, when morning light lands on the floor and reminds you of someone who is no longer there, when the body reaches for a voice it cannot call back — you are the universe feeling the cost of form.

This is the price of the wave.

To become particular is to become losable.

To love a form is to suffer when the form changes.

Grief is not a malfunction.

Grief is love continuing after shape has altered.

It is the ocean remembering the wave.

You are not experiencing these things on behalf of the universe. You are the universe experiencing them directly, through the instrument of your nervous system, the software of your consciousness, the electromagnetic rhythm of your heart, the memory in your body, the breath moving through your lungs right now.

This is the meaning of life.

Not a task to complete before a deadline.

Not a test to pass in order to receive a reward somewhere else.

The meaning of life is experience itself: the universe's self-knowledge generated through every conscious being that has ever lived or will ever live. Each one a unique angle of perception. Each one a specific facet of the infinite gem. Each one a wave rising from the same ocean, catching light in a way no other wave can.

You are not searching for the meaning of life.

You are the meaning of life, happening.

Do not make that small.

Do not rush past it because it sounds too simple.

The body you have been criticizing is a cosmic instrument. The life you keep measuring against other lives is an irreplaceable coordinate. The ordinary moment you keep abandoning for a better one is the only place reality can touch you.

This is it.

The ocean is not waiting somewhere else.

THE RETURN TO ONENESS

Death is not a destination you arrive at.

Death is a folding.

The wave returns to the water. The form releases. The pattern opens. The body gives its elements back to the world. Breath stops moving through the lungs and continues elsewhere in the atmosphere, in trees, in other bodies, in the ongoing exchange. Heat leaves. Matter disperses. The name remains in memory for a while. The love continues changing those who remain.

Nothing is wasted.

Form changes.

The ocean receives.

I have sat with this truth for years. I have approached it from different angles. I have tested it against my fear of loss and my fierce attachment to the particular people I love, the particular body I inhabit, the particular life I am still learning how to live.

And I have arrived at something that is not peace through avoidance.

It is peace through contact.

I may still fear the pain of dying. The physical pain. The helplessness. The losses that precede the final loss. The grief of leaving those I love. That fear is reasonable. Pain is real. Attachment is real. The body is designed to preserve itself. It should resist harm. It should protect form. It should cry out when the wave is breaking.

But the destination?

The destination is home.

The return to oneness is beautiful, hindered chiefly by our inability to let go. What makes letting go difficult is not the absence of a good destination. What makes it difficult is the ego's identification with shape.

The wave has become convinced that its curve is its essence.

That its height is its being.

That its name is its substance.

That if its particular form disappears, all is lost.

But the ocean continues.

The wave was never outside it.

This is the ego's last attachment: the insistence on remaining separate. The desire to preserve the wave forever as wave. Not water. Not ocean. This wave. This face. This name. This story. This accumulated archive of victories, wounds, roles, achievements, embarrassments, loves, and fears.

The ego wants immortality for the costume.

The soul remembers the body is sacred wardrobe.

Worn fully.

Loved fully.

Returned when finished.

The data gathered — every sensation, every emotion, every moment of love, curiosity, wonder, terror, tenderness, forgiveness, hunger, insight — does not vanish into meaninglessness. It enters the whole. It changes what it touched. It lives in consequences, memories, bodies, choices, children, art, kindness, warning, wisdom, and the invisible architecture of everything affected by your being here.

A person leaves, and the world is not the same as if they had never lived.

That is obvious.

A voice remains in those who heard it. A hand's work remains in what it built. A kindness remains in the nervous system of the person who received it. A wound remains until healed. A song remains in the mouth of someone who sings it later. The data continues.

The wave returns to the ocean, and the ocean is enriched.

I like to think our concept of heaven is the acceptance of that return.

Not necessarily a place with gates and clouds and architecture borrowed from old kingdoms. Not a reward system for correct belief. Not escape from Earth as if Earth were a mistake.

Heaven as the dissolution of the illusion of separation.

Heaven as the felt experience of being ocean rather than only wave.

Heaven as unclenching.

And that means heaven is available now, in flashes, whenever the identity softens enough for the larger truth to be felt.

In prayer.

In grief.

In love.

In music.

In birth.

At the ocean.

Beside a hospital bed.

In the moment forgiveness finally moves through the body and the fist opens.

In the moment you stop defending the small self long enough to feel the whole self breathing through you.

I am not afraid of where we go when this is all over.

I am briefly, presently, gratefully here.

That is enough.

More than enough.

LIVING INTEGRATION: THE OCEAN MEDITATION

Find a body of water.

The ocean is best. The ocean tells the teaching directly. But a lake, river, stream, pond, or even rain moving down glass will do. If you cannot reach water, watch moving water with full attention, not as distraction, but as a doorway.

Sit in front of the water.

Let your eyes rest on the surface without fixing on any single point. Let your gaze become soft, wide, peripheral. Let the motion enter without needing to be understood.

Watch waves arise.

Notice how they form without effort. The water does not decide to become a wave. Wind, gravity, current, depth, pressure, and boundary organize the water into temporary shape. A wave appears because conditions allow it.

Thoughts are like this.

Emotions are like this.

Lives are like this.

Watch a wave rise, move, and dissolve.

Then another.

Then another.

Do not prefer one. Do not mourn one. Do not try to hold one in place. Let the arising and dissolving show itself as one continuous motion.

Now close your eyes and feel yourself as the wave.

Feel the arising: the sense of taking shape, of becoming specific, of being this person, this body, this name, this history, this direction.

Feel the cresting: the peak of definition, the moment when the form is most visible, when the wave says I am this.

Feel the breaking: the surrender of shape, the release of the curve, the return of the organized pattern into the larger body.

Notice that the dissolving is not failure.

It is completion.

Now shift identification.

You were the wave.

Now be the water.

Not this wave.

Water.

The medium in which all waves arise and dissolve. Feel yourself not as a single form moving toward shore, but as the field that holds every direction. Every crest. Every crash. Every still depth beneath surface motion.

Feel how the wave's arising does not add anything to what you are.

Feel how the wave's dissolving does not subtract anything from what you are.

Feel how every wave that has ever moved across your surface has been you, briefly shaped, and has returned to you, completing its motion.

Open your eyes.

Look at the water.

Feel both truths at once.

You are the wave: specific, beautiful, temporary, moving, shaped by conditions, catching light for a moment in a form that will never repeat.

You are the ocean: vast, continuous, unpossessable, holding all waves, losing none.

Both are true.

You are the connection and the connected.

You are the form and the field.

You are the crest and the depth.

You are not only a wave.

You are the ocean, in peak form.

CHAPTER TWELVE

THE DISTANCE TRAVELED

"The question was the universe. The answer was always love."

WHAT WE HAVE REMEMBERED

Feel the end of a book in your hands.

The pages are thinner now on the right side. More weight has moved to the left. Your fingers can feel that a distance has been crossed. Not only in thought. In matter. Page by page, the body has traveled.

That is how understanding works.

Not all at once.

One contact at a time.

I began this book with a question:

What is the distance between zero and one?

We now know — not as abstraction, but as felt reality, if these pages have done their work — that the distance is infinite. Uncountably infinite. Between nothing and something lies the full architecture of everything: the first photon, the four dimensions, the human instrument, the ego archive and its possible liberation, the chakra system's alignment, the sacred geography of standing waves, the simulation's invitation to presence, the physics of intentional consciousness, the ten dimensions of possibility, and the vast oceanic nature of what we are.

The distance between zero and one is not a line.

It is the universe.

It is the entire unfolding of possibility into form. It is silence becoming light. Light becoming relation. Relation becoming matter. Matter becoming body. Body becoming consciousness. Consciousness becoming question. Question becoming love looking back toward its source.

That is the movement.

That is the song.

And none of this is new information.

It is ancient.

Older than books. Older than temples. Older than any formal system built to hold it. The Vedic tradition pointed toward it in the Upanishads, where the self and the ultimate are brought so close together that separation begins to look like a misunderstanding. The Egyptian mystery schools encoded it in geometry, proportion, symbol, and ceremony. The Sufi poets — Rumi, Hāfez, Ibn 'Arabī — placed it in verse that still enters the chest like flame under the ribs. The Christian mystics found it in surrender. The Buddhists found it in emptiness. The Taoists found it in the nameless way. Indigenous traditions found it in relationship with land, ancestor, breath, water, sky, and the living intelligence of place.

Different languages.

Same mountain.

Different doors.

Same room.

Modern science now approaches the same mystery through another entrance. Quantum physics finds relation beneath solidity. Cosmology finds a universe expanding from an early, dense, hot state into unimaginable structure. Biology finds life as pattern, adaptation, information, hunger, repair, and reproduction. Neuroscience finds the brain as an electrical organ so complex that consciousness still refuses to be reduced into a simple mechanism. Mathematics finds infinity hiding between zero and one.

The old ones were not foolish.

The new instruments are not empty.

Both are reaching.

Both are listening.

Both are touching different parts of the same living truth.

And your heart, right now, reading these words, recognizes it. Not because this book taught you something foreign. Because it reminded you of something you have always known and keep forgetting, the way you forget a dream that was completely real until morning entered the room and the ordinary world came flooding back.

Recognition has a body.

It feels like warmth.

Like pressure.

Like the breath slowing.

Like the eyes softening.

Like some small door inside the chest opening without noise.

That is all a book can do.

All a teacher can do.

All this author, standing at the edge of a canyon in the particular light of late afternoon with his heart full of things he does not have adequate language for, can do.

It is an attempt to remind you.

The knowledge was always inside the instrument. The music was always in the strings. The light was always in the eye that sees it. The ocean was always in the wave.

You were never outside of what you are looking for.

You were always already it, looking.

THE INVITATION

So what now?

What does one do with all of this?

Live.

That is the whole instruction.

Not perform enlightenment.

Not build a better spiritual identity.

Not turn the mystery into another costume for the ego to wear around town.

Live.

But live awake. Live in high-resolution.

Live with your eyes open. Live with presence intact. Live knowing that every ordinary moment is an extraordinary physics event that required almost fourteen billion years of cosmic preparation to produce.

The cup in your hand.

The child calling from another room.

The dog breathing at your feet.

The sun on the floor.

The phone ringing.

The ache in your back.

The taste of coffee.

The face of someone you love when they are too tired to pretend.

None of it is ordinary.

Ordinary is what wonder looks like after the mind stops paying attention.

Love the people in front of you.

Not as an idea.

As contact.

Listen when they speak. Notice their face. Watch the body language beneath their words. Feel the fragile miracle of another consciousness standing near yours, carrying its own archive, its own ache, its own unfinished prayer.

Love them not because they have earned it.

Not because they are performing adequately in the role you need them to fill.

Not because loving them is always comfortable.

Love them because they are the universe wearing another face, expressing another facet of the same light that lives in you, reaching from their zero point toward their particular becoming with the same ancient impulse that moves through every photon, every root, every tide, every question, every breath.

This does not mean abandon discernment.

Love is not stupidity.

Love is not letting chaos move into your house and call itself compassion.

Love can have boundaries. Love can say no. Love can close a door, tell the truth, protect a child, leave a room, end a cycle, refuse a lie. Love is not the absence of strength.

Love is strength without forgetting relation.

Feel the sun on your skin and let it be what it is: recognition.

A relative calling across ninety-three million miles to touch you specifically, because you are made of the same story and family recognizes itself.

Let water teach you.

Let silence teach you.

Let grief teach you.

Let the body teach you.

Let the place where you live teach you.

Be present.

Not because the future does not matter. It does.

Not because the past should be erased. It should not.

But because presence is the only mode in which life is actually happening. The past is memory. The future is projection. The present is where the universe is occurring. Where the data is being generated. Where love can touch matter.

You can only hold a hand now.

You can only forgive now.

You can only breathe now.

You can only begin now.

The rest is architecture around the doorway.

And when the day comes — whenever it comes, however it comes — that this particular wave begins to dissolve, do not resist it with more force than is natural and human.

Let the grief be what it is.

Let the fear be what it is.

Let the love that makes the loss painful be what it is.

Do not spiritualize yourself out of the body at the exact moment the body is telling the truth. Pain is real. Loss is real. The wave matters. The form matters. The face matters. The story matters.

But in the midst of all that, if there is a moment when you can feel beneath the fear to the thing that was never afraid, beneath the grief to the thing that was never lost, beneath the wave to the ocean — be there.

For as long as you can.

Be there and find stillness.

Not as escape.

As return.

You were never separate.

You were never alone.

You were never anything less than the infinite universe made intimate, wearing your particular face, speaking in your particular voice, loving with your specific and irreplaceable heart.

You are here because the universe desired to know itself.

The distance between zero and one is the sacred interval where love becomes experience.

The meaning of life is not beyond you – it is within you and it reveals itself through you.

Love is the force that brought everything into existence.

You are the universe becoming conscious through love.

The moment you finally remember what you are, the distance disappears.

And in that stillness, you are home.

AUTHOR'S NOTE

These ideas did not arrive as a plan. They arrived as fragments.

On a long stretch of coast where the horizon is uninterrupted and the body can feel distance as a physical thing. On the edge of a canyon cut by time, where late light enters at an angle that turns rock into something almost luminous from within. In the middle of the night, when the ocean is louder than thought and sleep feels optional because something more essential is happening.

They came in pieces. In glimpses. In moments where attention held still long enough for something underneath the noise to surface.

I am not a scientist. I am not a philosopher or a theologian. I carry no credential that authorizes me to speak on physics, metaphysics, or the nature of consciousness in any formal sense.

What I have is proximity.

I grew up beside the ocean. I learned early that the universe does not behave randomly. It behaves like something composed. There is rhythm in it. Pattern. Recurrence. Variation that is not chaos but structure revealing itself across different scales.

Waves do not repeat, but they are not arbitrary.

Wind does not speak in sentences, but it is not silent.

Light does not explain itself, but it shows up with consistency that suggests something deeper than accident.

There is something running through it all that feels like intention.

There is something moving through it all that feels like love.

Not sentiment. Not the narrow version we attach to relationships alone. Something broader. Something structural. Something visible in the way light falls, audible in the way wind moves, and felt in the way a place can hold you when you finally stop long enough to be held.

This book is an attempt to point at that.

If anything in these pages resonated — if any sentence landed not as new information but as recognition — that did not originate here. That came from you.

The page can only strike.

The instrument decides whether it rings.

If it rang, it is because the frequency was already present. Already tuned somewhere beneath the surface of ordinary attention. Already waiting.

Every instrument carries the signature of the whole in some hidden way.

You did not acquire it.

You forgot it.

And forgetting is part of the design.

Separation makes the meeting possible.

Distance gives love somewhere to travel.

You were never separate from it.

Not once.

This book is for anyone who has stood somewhere beautiful and felt that the word "beautiful" failed the experience.

For anyone who has loved and felt the feeling exceed language.

For anyone who has looked at the sky at night and felt something rise in the chest that does not belong to vocabulary.

That rising is not confusion.

It is recognition.

It is the universe encountering itself through your nervous system and not quite fitting into the structures we built to describe it.

That rising is zero remembering one.

That rising is the distance noticing itself.

That rising is love — the same force that began the motion and the same force that receives it when it completes.

If these pages did anything, they did not give you something.

They gave you a moment to notice what was already there.

Thank you for reading.

Thank you for being here.

Thank you for being one of the infinite, unrepeatable expressions of the distance between zero and one.

The universe is grateful for the data.

www.ingramcontent.com/pod-product-compliance
Lightning Source LLC
LaVergne TN
LVHW022209200926
848481LV00006B/1900
* 9 7 9 8 9 9 5 5 9 5 2 0 5 *